Encyclopaedia of Life Science

Encyclopaedia of Life Science

VOL. 1

B.A. Mannan

ANMOL PUBLICATIONS PVT. LTD.

NEW DELHI - 110 002 (INDIA)

ANMOL PUBLICATIONS PVT. LTD.
H.O.: 4374/4B, Ansari Road, Darya Ganj,
New Delhi-110 002 (India)
Ph.: 23278000, 23261597
B.O.: No. 1015, Ist Main Road, BSK IIIrd Stage
IIIrd Phase, IIIrd Block,
Bangalore - 560 085 (India)
Visit us at: www.anmolpublications.com

Encyclopaedia of Life Science

First Published, 2007
ISBN 978-81-261-3408-3 (Set)

PRINTED IN INDIA
Printed at Amit Enterprises, Delhi.

Contents

Preface

'Life Science' is a branch of science, under which we study about the living beings in detail. We study about the. body structure; all the organs of the body and their functions. The body consists of several organs, which are made of Tissues, while, Tissues are made of cells. Plants are very useful for living organisms as they supply oxygen for the respiration and they are the food-source for the primary consumers. Plant survival and growth is critically influenced by abiotic factors including water, air and light. The effect of the local environment on plant growth also accounts for much of variation in adult form within a species. Plants have green pigments, which are quite useful in the process of photosynthesis. In the presence of light, plants make their food with the help of minerals, water and carbondioxide, in their green leaves, in the form of carbohydrates. The food prepared in green leaves is transported to different parts of the plant by phloem.

Similarly, animal life is a significant feature on earth. All the functions of the human beings and animals, including birds and insects, are performed in their cells. Cell is the structural unit of any living body. This was first discovered by scientist, Robert Hook. The cell has several parts, such as nucleus, centrosomes, mitochondria, plastids, cell-walls. The cell wall gives the mechanical support to the cell and cell-walls are accountable for the transportation of the material in and out of the cells.

Biologists agree that all living things arose through long periods of change shaped by physical and chemical processes. Approximately, there are two million species in the living world. In changing environment, a few species are eliminated during un-favourable conditions on the basis of the principle of the 'survival of the fittest,' which is also known as natural selection.

Science can strive to grasp nature's faculty, but science is socially embedded and all scientists record prevailing certainties. Darwin himself, in the closing lines of 'The Origin of Species', mentioned about Victorial social preference. Life's pathway certainly includes many features, predicable from laws of nature. But laws of nature do not tell us why we have crabs and snails at all, why insects rule the multicellular world, and why vertebrates rather than persistent algal mats exist as the most complex form of life on earth.

Bacteria too represent the great success story of life's pathway. They occupy a wider domain of environments and span a border range of biochemistries than any other group. One might conclude that complexification of life as a whole represents a pseudo-trend based on constraints at the wall, but still holds that evaluation continues, within a particular group.

This eminent work, titled; *Encyclopaedia of Life Science* would be able to make new waves in the discipline. Expectedly, these books would prove to be the best source of study in the area of Life Science.

Editor

1

Introduction

Evolution is the process by which all living things have developed from primitive organisms through changes occurring over billions of years, a progression that includes the most advanced animals and plants. Exactly how evolution occurs is still a matter of debate, but that it occurs is a scientific fact.

Biologists agree that all living things arose through a long history of changes shaped by physical and chemical processes that are still taking place. It is plausible that all organisms can be traced back to the origin of Life from inanimate matter. The most direct proof of evolution is furnished by the science of Palaeontology, or the study of life in the past through fossil remains or impressions, usually in rock. Additional evidence comes from comparative studies of living animals and plants, including their structure, Biochemistry Iochemistry, Embriology, and geographical distribution.

Approximately 2 million different species of organisms are now living, but it is estimated that at least 99.9 per cent of the species that have ever lived are now extinct and that some 2 billion species have evolved during the past 600 million years. Changes occur in living organisms that serve to increase their adaptability, or potential for survival and reproduction, in the face of changing environments. Evolution apparently has no built-in direction or foreordained purpose. A given kind of organism may evolve only when it occurs in a variety of forms differing in hereditary characteristics, or traits, that are passed from parent to offspring. Purely by chance, some varieties prove to be ill adapted to their current environment and thus disappear, whereas others prove to be adaptive, and their numbers increase.

The elimination of the unfit, or the "survival of the fittest," is known as Natural selection because it is nature that discards or favours a particular variant. Basically, evolution takes place only when natural selection operates on a population of organisms containing diverse inheritable forms. Recently, natural selection was demonstrated for the first time outside of the laboratory when scientists observed guppies change their reproductive behaviour over an 11-year period in direct response to being transferred to a new environment that had different predators.

Evolution is a revolutionary concept that took the human world by surprise when it was first proposed by Darwin almost 150 years ago, encountering a heavy amount of criticism and speculation that hasn't been able to deny the feasibility of this process taking place. The theory of evolution has survived the great amount of change experienced in our modern time of swift advances in science & technology which has only modified it slightly and maintained it's existence as a valid and reasonable theory, allowing it to hold up remarkably well under the intense and unrelenting scrutiny and testing of the scientific community.

Haven't you ever wondered about how life came to be? Why is it so varied ? Why do some living beings seem to look more alike than others ? These are all questions to which evolution finds a reasonable and scientific answer supported by tons of evidence. It is a process of gradual change that substantiates how life arose from non- living matter and slowly developed throughout millions of years into diverse groups of living creatures that are in fact related due to a common ancestor, which have and will continue to develop into other species.

Humans aren't and never will be the uttermost achievement of the evolutionary process. "And on the first day God created..." Man, in his desperate search for answers, attempted to find an explanation to how he and his world came to exist. This was the motivation behind the many myths of creation that prevailed in human religions and cultures, of which the most remarkable and famous is the Biblical account; which is simply that, a majestic and enlightening explanation that uses a creator's power as a backdrop and is not supported by any reasonable evidence.

Many religious believers firmly reject the evolutionary theory because of the faith they have in a belief that has been the foundation behind their perspective of life, but it is simply a tale created by a human group that tried to satisfy it's search for answers and came to the point were they had to make them up. Science, in contrast, is in search of the ultimate and most reasonable explanation supported by the evidence, where beliefs are not an influence in the search for answers. This is the only way to obtain an objective explanation that is adept to reality.

Great revolutions in the history of science have but one common, and ironic, feature: they knock human arrogance off one pedestal after another of our previous conviction about our own self-importance. In Freud's three examples, Copernicus moved our home from centre to periphery, Darwin then relegated us to "descent from an animal world"; and, finally (in one of the least modest statements of intellectual history).

In this wise and crucial sense, the Darwinian revolution remains woefully incomplete because, even though thinking humanity accepts the fact of evolution, most of us are still unwilling to abandon the comforting view that evolution means (or at least embodies a central principle of) progress defined to render the appearance of something like human consciousness either virtually inevitable or at least predictable.

The pedestal is not smashed until we abandon progress or complexification as a central principle and come to entertain the strong possibility that H. sapiens is but a tiny, late-arising twig on life's enormously arborescent bush - a small bud that would almost surely not appear a second time if we could replant the bush from seed and let it grow again. The veracity of the theory of evolution is evident when you compare it with the technological application of other laws and theories.

Electrons or their equivalent must exist, else our machines based on the theories and ideas about electromagnetism wouldn't work. We know gravity exists because we're not floating all over the place, and so the laws of gravity maintain their standing when applied to the real world. By the same token evolution must exist, or else genetic engineering wouldn't work and we would be incapable of cloning and manipulating living beings, things which have been recently accomplished and apply the concepts of evolution, thereby obtaining results.

Everything we expect in genomes, populations, lab experience, ecology, palaeontology, all fit the model great! For the moment, we consider it to be the unifying idea in biology...in that way, it is true but true only in the loosest sense of the word! No, a theory is not something that would be a fact or a law if only there were enough supporting evidence for it.

A theory is a detailed explanation of how and/or why certain phenomena occur, usually in terms of more basic phenomena; it is neither a fact nor a law. Creationists have polluted the scientific meaning of the term "theory" by squawking that "evolution is only a theory, not a fact." But comparing theories with facts in this manner is like comparing apple trees with apples: although the objects are related, the comparison is pointless. For example, the theory of relativity explains how and why time slows down for moving objects (an observed fact). Likewise, the theory of evolution by natural selection explains why earth's life forms have changed through time (another observed fact). Theories may be modified or even overturned, but only in favour of a better or deeper or broader explanation of a body of observed facts.

Something that may or may not be important here: natural selection is just a proposed mechanism of evolution. All the evidence we have indicates that organisms have changed through time, and we know through observation that the allele frequencies of a population can change over time. This would indicate that the statement "Evolution does occur" is a fact. Proposing that it occurs via natural selection is a theory. Primates are visual animals, and the pictures we draw betray our deepest convictions and display our current conceptual limitations.

An error that artists have always made in the portrayal of evolution is painting the history of fossil life as a sequence from invertebrates, to fishes, to early terrestrial amphibians and reptiles, to dinosaurs, to mammals and, finally, to humans. There are no exceptions; all sequences painted since the inception of this genre in the 1850s follow the convention.

Yet we never stop to recognise the almost absurd biases coded into this universal mode and which we need to and have come to change in the last couple of years. No scene ever shows another invertebrate after fishes evolved but invertebrates did not go away or stop evolving! After terrestrial reptiles emerge, no subsequent scene ever shows a fish (later oceanic tableaux depict only such returning reptiles as ichthyosaurs and plesiosaurs). But fishes did not stop evolving after one small lineage managed to invade the land. In fact, the major event in the evolution of fishes, the origin and rise to dominance of the teleosts, or modern bony fishes, occurred during the time of the dinosaurs and is therefore never shown at all in any of these sequences -even though teleosts include more than half of all species of vertebrates.

Why should humans appear at the end of all sequences? Our order of primates is ancient among mammals, and many other successful lineages arose later than we did. We will not smash Freud's pedestal and complete Darwin's revolution until we find, grasp and accept another way of drawing life's history. J.B.S. Haldane proclaimed nature "queerer than we can suppose," but these limits may only be socially imposed conceptual locks rather then inherent restrictions of our neurology. New icons might break the locks. Trees - or rather copiously and luxuriantly branching bushes - rather than ladders and sequences hold the key to this conceptual transition.

We must learn to depict the full range of variation, not just our parochial perception of the tiny right tail of most complex creatures. We must recognise that this tree may have contained a maximal number of branches near the beginning of multicellular life and that subsequent history is for the most part a process of elimination and lucky survivorship of a few, rather than continuous flowering, progress and expansion of a growing multitude. We must understand that little twigs are contingent nubbins, not predictable goals of the massive bush beneath.

In our time the theory of evolution has come to be a fact and the central concept of biology, it is certain that gradual changes took place in living beings and that they are all related. What still isn't positively known is how it takes place, is natural selection the answer ?, what other methods could the process take place with? The only new ideas that have changed about evolution is the notion that it is guided to a subsequently higher complexity, but instead the survival of the fittest that sometimes might need to obtain superiority over other creatures by adapting properly to it's environment, complexity mostly ensues over competitors but it isn't always the answer.

Indeed, you only need to stop and think about how lucky you are to form part of a species that has come to explore the concept of it's own origin and mysterious development. The only question know left to answer is the universality of evolution. If life is ever found in other planets (which it surely will) we will come to question ourselves if evolution is also responsible for the development of the alien being and if our common ancestor is non- living matter, in essence the Big Bang that originated time and the universe.

The history of life is not necessarily progressive; it is certainly not predictable. The earth's creatures have evolved through a series of ccntingent and fortuitous events. Some creators announce their inventions with grand eclat. God proclaimed, "Fiat lux," and then flooded his new universe with brightness. Others bring forth great discoveries in a modest guise, as did Charles Darwin in defining his new mechanism of evolutionary causality in 1859: "this principle have called, by which each slight variation, if useful, is preserved, by the term Natural Selection."

Natural selection is an immensely powerful yet beautifully simple theory that has held up remarkably well, under intense and unrelenting scrutiny and testing, for 135 years. In essence, natural selection locates the mechanism of evolutionary change in a "struggle" among organisms for reproductive success, leading to improved fit of populations to changing environments. (Struggle is often a metaphorical description and need not be viewed as overt combat, guns blazing. Tactics for reproductive success include a variety of non-martial activities such as earlier and more frequent mating or better cooperation with partners in raising offspring.) Natural selection is therefore a principle of local adaptation, not of general advance or progress.

Slab containing specimens of Pteridinium from Namibia shows a prominent organism from thc earth's first multicellular fauna, called Ediacaran, which appeared some 600 million years ago. The Ediacaran animals died out before the Cambrian explosion of modern life.

These thin, quilted, sheetlike organisms may be ancestral to some modern forms but may also represent a separate and ultimately failed experiment in multicellular life. The history of life tends to move in quick and quirky episodes, rather than by gradual improvement.

Yet powerful though the principle may be, natural selection is not the only cause of evolutionary change (and may, in many cases, be overshadowed by other forces). This point needs emphasis because the standard misapplication of evolutionary theory assumes that biological explanation may be equated with devising accounts, often speculative and conjectural in practice, about the adaptive value of any given feature in its original environment (human aggression as good for hunting, music and religion as good for tribal cohesion, for example).

Darwin himself strongly emphasised the multifactorial nature of evolutionary change and warned against too exclusive a reliance on natural selection, by placing the following statement in a maximally conspicuous place at the very end of his introduction:

"I am convinced that Natural Selection has been the most important, but not the exclusive, means of modification."

Natural selection is not fully sufficient to explain evolutionary change for two major reasons. First, many other causes are powerful, particularly at levels of biological organisation both above and below the traditional Darwinian focus on organisms and their struggles for reproductive success. At the lowest level of substitution in individual base pairs of DNA, change is often effectively neutral and therefore random. At higher levels, involving entire species or faunas, punctuated equilibrium can produce evolutionary trends by selection of species based on their rates of origin and extirpation, whereas mass extinctions wipe out substantial parts of biotas for reasons unrelated to adaptive struggles of constituent species in "normal" times between such events.

Theory, of course, is relevant to explaining the pathway (nothing about the pathway can be inconsistent with good theory, and theory can predict certain general aspects of life's geologic pattern). But the actual pathway is strongly underdetermined by our general theory of life's evolution. This point needs some belabouring as a central yet widely misunderstood aspect of the world's complexity. Webs and chains of historical events are so intricate, so imbued with random and chaotic elements, so unrepeatable in encompassing such a multitude of unique (and uniquely interacting) objects, that standard models of simple prediction and replication do not apply.

History can be explained, with satisfying rigor if evidence be adequate, after a sequence of events unfolds, but it cannot be predicted with any precision beforehand. Pierre-Simon Laplace, echoing the growing and confident determinism of the late 18th century, once said that he could specify all future states if he could know the position and motion of all particles in the cosmos at any moment, but the nature of universal complexity shatters this chimerical dream. History includes too much chaos, or extremely sensitive dependence on minute and unmeasurable differences in initial conditions, leading to massively divergent outcomes based on tiny and unknowable disparities in starting points.

Progress does not rule (and is not even a primary thrust of) the evolutionary process. For reasons of chemistry and physics, life arises next to the "left wall" of its simplest conceivable and preservable complexity. This style of life (bacterial) has remained most common and most successful. A few creatures occasionally move to the right, thus extending the right tail in the distribution of complexity. Many always move to the left, but they are absorbed within space already occupied.

And history includes too much contingency, or shaping of present results by long chains of unpredictable antecedent states, rather than immediate determination by timeless laws of nature. Homo sapiens did not appear on the earth, just a geologic second ago, because evolutionary theory predicts such an outcome based on themes of progress and increasing neural complexity. Humans arose, rather, as a fortuitous and contingent outcome of thousands of linked events, any one of which could have occurred differently and sent history on an alternative pathway that would not have led to consciousness. To cite just four among a multitude:

(1) If our inconspicuous and fragile lineage had not been among the few survivors of the initial radiation of multicellular animal life in the Cambrian explosion 530 million years ago, then no vertebrates would have inhabited the earth at all. (Only one member of our chordate phylum, the genus Pikaia, has been found among these earliest fossils. This small and simple swimming creature, showing its allegiance to us by possessing a notochord, or dorsal stiffening rod, is among the rarest fossils of the Burgess Shale, our best preserved Cambrian fauna.)

(2) If a small and unpromising group of lobe-finned fishes had not evolved fin bones with a strong central axis capable of bearing weight on land, then vertebrates might never have become terrestrial.

(3) If a large extraterrestrial body had not struck the earth 65 million years ago, then dinosaurs would still be dominant and mammals insignificant (the situation that had prevailed for 100 million years previously).

(4) If a small lineage of primates had not evolved upright posture on the drying African savannas just two to four million years ago, then our ancestry might have ended in a line of apes that, like the chimpanzee and gorilla today, would have become ecologically marginal and probably doomed to extinction despite their remarkable behavioural complexity.

Therefore, to understand the events and generalities of life's pathway, we must go beyond principles of evolutionary theory to a palaeontological examination of the contingent pattern of life's history on our planet - the single actualised version among millions of plausible alternatives that happened not to occur. Such a view of life's history is highly contrary both to conventional deterministic models of Western science and to the deepest social traditions and psychological hopes of Western culture for a history culminating in humans as life's highest expression and intended planetary steward.

Science can, and does, strive to grasp nature's factuality, but all science is socially embedded, and all scientists record prevailing "certainties," however hard they may be aiming for pure objectivity. Darwin himself, in the closing lines of The Origin of Species, expressed Victorian social preference more than nature's record in writing: "As natural selection works solely by and for the good of each being, all corporeal and mental endowments will tend to progress towards perfection." Life's pathway certainly includes many features predictable from laws of nature, but these aspects are too broad and general to provide the "Tightness" that we seek for validating evolution's particular results - roses, mushrooms, people and so forth.

Organisms adapt to, and are constrained by, physical principles. It is, for example, scarcely surprising, given laws of gravity, that the largest vertebrates in the sea (whales) exceed the heaviest animals on land (elephants today, dinosaurs in the past), which, in turn, are far bulkier than the largest vertebrate that ever flew (extinct pterosaurs of the Mesozoic era). Predictable ecological rules govern the structuring of communities by principles of energy flow and thermodynamics.

Evolutionary trends, once started, may have local predictability. But laws of nature do not tell us why we have crabs and snails at all, why insects rule the multicellular world and why vertebrates rather than persistent algal mats exist as the most complex forms of life on the earth. Relative to the conventional view of life's history as an at least broadly predictable process of gradually advancing complexity through time, three features of the palaeontological record stand out in opposition and shall therefore serve as organising themes: the constancy of modal complexity throughout life's history;

the concentration of major events in short bursts interspersed with long periods of relative stability; and the role of external impositions, primarily mass extinctions, in disrupting patterns of "normal" times. These three features, combined with more general themes of chaos and contingency, require a new framework for conceptualising and drawing life's history.

The primary palaeontological fact about life's beginnings points to predictability for the onset and very little for the particular pathways thereafter. The earth is 4.6 billion years old, but the oldest rocks date to about 3.9 billion years because the earth's surface became molten early in its history, a result of bombardment by large amounts of cosmic debris during the solar system's coalescence, and of heat generated by radioactive decay of short-lived isotopes. These oldest rocks are too metamorphosed by subsequent heat and pressure to preserve fossils (though some scientists interpret the proportions of carbon isotopes in these rocks as signs of organic production).

The oldest rocks sufficiently unaltered to retain cellular fossils - African and Australian sediments dated to 3.5 billion years old - do preserve prokaryotic cells (bacteria and cyanophytes) and stromatolites (mats of sediment trapped and bound by these cells in shallow marine waters). Thus, life on the earth evolved quickly and is as old as it could be. This fact alone seems to indicate an inevitability, or at least a predictability, for life's origin from the original chemical constituents of atmosphere and ocean.

No one can doubt that more complex creatures arose sequentially after this prokaryotic beginning - first eukaryotic cells, perhaps about two billion years ago, then multicellular animals about 600 million years ago, with a relay of highest complexity among animals passing from invertebrates, to marine vertebrates and, finally, to reptiles, mammals and humans. This is the conventional sequence represented in the old charts and texts as an "age of invertebrates," followed by an "age of fishes," "age of reptiles," "age of mammals," and "age of man". The most salient feature of life has been the stability of its bacterial mode from the beginning of the fossil record until today and, with little doubt, into all future time so long as the earth endures. This is truly the "age of bacteria"- as it was in the beginning, is now and ever shall be.

New Iconography of Life Tree

New iconography of life tree shows that maximal diversity in anatomical forms (not in number of species) is reached very early in life's multicellular history. Later times feature extinction of most of these initial experiments and enormous success within surviving lines. This success is measured in the proliferation of species but not in the development of new anatomies. Today we have more species than ever before, although they are restricted to fewer basic anatomies.

For reasons related to the chemistry of life's origin and the physics of self- organisation, the first living things arose at the lower limit of life's conceivable, preservable complexity. Call this lower limit the "left wall" for an architecture of complexity. Since so little space exists between the left wall and life's initial bacterial mode in the fossil record, only one direction for future increment exists—towards greater complexity at the right. Thus, every once in a while, a more complex creature evolves and extends the range of life's diversity in the only available direction. In technical terms, the distribution of complexity becomes more strongly right skewed through these occasional additions.

But the additions are rare and episodic. They do not even constitute an evolutionary series but form a motley sequence of distantly related taxa, usually depicted as eukaryotic cell, jellyfish, trilobite, nautiloid, eurypterid (a large relative of horseshoe crabs), fish, an amphibian such as Eryops, a dinosaur, a mammal and a human being. This sequence cannot be construed as the major thrust or trend of life's history. Think rather of an occasional creature tumbling into the empty right region of complexity's space. Throughout this entire time, the bacterial mode has grown in height and remained constant in position.

Bacteria represent the great success story of life's pathway. They occupy a wider domain of environments and span a broader range of biochemistries than any other group. They are adaptable, indestructible and astoundingly diverse. We cannot even imagine how anthropogenic intervention might threaten their extinction, although we worry about our impact on nearly every other form of life. The number of Escherichia coli cells in the gut of each human being exceeds the number of humans that has ever lived on this planet.

One might grant that complexification for life as a whole represents a pseudo-trend based on constraint at the left wall but still hold that evolution within particular groups differentially favours complexity when the founding lineage begins far enough from the left wall to permit movement in both directions. Empirical tests of this interesting hypothesis are just beginning (as concern for the subject mounts among paleontologists), and we do not yet have enough cases to advance a generality. Moreover, when we consider that for each mode of life involving greater complexity, there probably exists an equally advantageous style based on greater simplicity of form (as often found in parasites, for example), then preferential evolution towards complexity seems unlikely a priori.

Our impression that life evolves towards greater complexity is probably only a bias inspired by parochial focus on ourselves, and consequent over-attention to complexifying creatures, while we ignore just as many lineages adapting equally well by becoming simpler in form. The morphologically degenerate parasite, safe within its host, has just as much prospect for evolutionary success as its gorgeously elaborate relative coping with the slings and arrows of outrageous fortune in a tough external world.

Even if complexity is only a drift away from a constraining left wall, we might view trends in this direction as more predictable and characteristic of life's pathway as a whole if increments of complexity accrued in a persistent and gradually accumulating manner through time. But nothing about life's history is more peculiar with to this common (and false) respect expectation than the actual pattern of extended stability and rapid episodic movement, as revealed by the fossil record.

Life remained almost exclusively unicellular for the first five sixths of its history - from the first recorded fossils at 3.5 billion years to the first well- documented multicellular animals less than 600 million years ago. (Some simple multicellular algae evolved more than a billion years ago, but these organisms belong to the plant kingdom and have no genealogical connection with animals.) This long period of unicellular life does include, to be sure, the vitally important transition from simple prokaryotic cells without organelles to eukaryotic cells with nuclei, mitochondria and other complexities of intracellular architecture—but no recorded attainment of multicellular animal organisation for a full three billion years.

If complexity is such a good thing, and multicellularity represents its initial phase in our usual view, then life certainly took its time in making this crucial step. Such delays speak strongly against general progress as the major theme of life's history, even if they can be plausibly explained by lack of sufficient atmospheric oxygen for most of Pre-cambrian time or by failure of unicellular life to achieve some structural threshold acting as a prerequisite to multicellularity. More curiously, all major stages in organising animal life's multicellular architecture then occurred in a short period beginning less than 600 million years ago and ending by about 530 million years ago - and the steps within this sequence are also discontinuous and episodic, not gradually accumulative.

The first fauna, called Ediacaran to honour the Australian locality of its initial discovery but now known from rocks on all continents, consists of highly flattened fronds, sheets and circlets composed of numerous slender segments quilted together. The nature of the Ediacaran fauna is now a subject of intense discussion. These creatures do not seem to be simple precursors of later forms. They may constitute a separate and failed experiment in animal life, or they may represent a full range of diploblastic (two-layered) organisation, of which the modern phylum Cnidaria (corals, jellyfishes and their allies) remains as a small and much altered remnant.

In any case, they apparently died out well before the Cambrian biota evolved. The Cambrian then began with an assemblage of bits and pieces, frustratingly difficult to interpret, called the "small shelly fauna." The subsequent main pulse, starting about 530 million years ago, constitutes the famous Cambrian explosion, during which all but one modern phylum of animal life made a first appearance in the fossil record. The Bryozoa, a group of sessile and colonial marine organisms, do not arise until the beginning of the subsequent, Ordovician period, but this apparent delay may be an artifact of failure to discover Cambrian representatives.

Great diversity quickly evolved at the dawn of multicellular animal life during the Cambrian period (530 million years ago). The creatures shown here are all found in the Middle Cambrian Burgess Shale fauna of Canada. They include some familiar forms (sponges, brachiopods) that have survived. But many creatures (such as the giant Anomalocaris at the lower right, largest of all the Cambrian animals) did not live for long and are so anatomically peculiar (relative to survivors) that we cannot classify them among known phyla.

Although interesting and portentous events have occurred since, from the flowering of dinosaurs to the origin of human consciousness, we do not exaggerate greatly in stating that the subsequent history of animal life amounts to little more than variations on anatomical themes established during the Cambrian explosion within five million years. Three billion years of unicellularity, followed by five million years of intense creativity and then capped by more than 500 million years of variation on set anatomical themes can scarcely be read as a predictable, inexorable or continuous trend towards progress or increasing complexity.

We do not know why the Cambrian explosion could establish all major anatomical designs so quickly. An "external" explanation based on ecology seems attractive: the Cambrian explosion represents an initial filling of the "ecological barrel" of niches for multicellular organisms, and any experiment found a space. The barrel has never emptied since; even the great mass extinctions left a few species in each principal role, and their occupation of ecological space forecloses opportunity for fundamental novelties. But an "internal" explanation based on genetics and development also seems necessary as a complement: the earliest multicellular animals may have maintained a flexibility for genetic change and embryological transformation that became greatly reduced as organisms "locked in" to a set of stable and successful designs.

In any case, this initial period of both internal and external flexibility yielded a range of invertebrate anatomies that may have exceeded (in just a few million years of production) the full scope of animal form in all the earth's environments today (after more than 500 million years of additional time for further expansion). Scientists are divided on this question. Some claim that the anatomical range of this initial explosion exceeded that of modern life, as many early experiments died out and no new phyla have ever arisen. But scientists most strongly opposed to this view allow that Cambrian diversity at least equalled the modern range - so even the most cautious opinion holds that 500 million subsequent years of opportunity have not expanded the Cambrian range, achieved in just five million years.

The Cambrian explosion was the most remarkable and puzzling event in the history of life. Moreover, we do not know why most of the early experiments died, while a few survived to become our modern phyla. It is tempting to say that the victors won by virtue of greater anatomical complexity, better ecological fit or some other predictable feature of conventional Darwinian struggle. But no recognised traits unite the victors, and the radical alternative must be entertained that each early experiment received little more than the equivalent of a ticket in the largest lottery ever played out on our planet—and that each surviving lineage, including our own phylum of vertebrates, inhabits the earth today more by the luck of the draw than by any predictable struggle for existence.

The history of multicellular animal life may be more a story of great reduction in initial possibilities, with stabilisation of lucky survivors, than a conventional tale of steady ecological expansion and morphological progress in complexity. Finally, this pattern of long stasis, with change concentrated in rapid episodes that establish new equilibria, may be quite general at several scales of time and magnitude, forming a kind of fractal pattern in self-similarity. According to the punctuated equilibrium model of speciation, trends within lineages occur by accumulated episodes of geologically instantaneous speciation, rather than by gradual change within continuous populations (like climbing a staircase rather than rolling a ball up an inclined plane).

Even if evolutionary theory implied a potential internal direction for life's pathway, the occasional imposition of a rapid and substantial, perhaps even truly catastrophic, change in environment would have intervened to stymie the pattern.

History of Life and Classical Representation

Life's history and classical representation reveal the severe biases of viewing evolution as embodying a central principle of progress and complexification. In these paintings, the first panel shows invertebrates of the Burgess Shale. But as soon as fishes evolve (panel 2), no subsequent scene ever shows another invertebrate, although they did not go away or stop evolving. When land vertebrates arise (panel 3), we never see another fish, even though return of land vertebrate lineages to the sea may be depicted (panel 4). The sequence always ends with mammals (panel 5)-even though fishes, invertebrates and reptiles are still thriving-and, of course, humans (panel 6).

Species and may so derail any internal direction and so reset the pathway that the net pattern of life's history looks more capricious and concentrated in episodes than steady and directional. Mass extinctions have been recognised since the dawn of palaeontology; the major divisions of the geologic time scale were established at boundaries marked by such events. But until the revival of interest that began in the late 1970S, most paleontologists treated mass extinctions only as intensifications of ordinary events, leading (at most) to a speeding up of tendencies that pervaded normal times.

In this gradualistic theory of mass extinction, these events really took a few million years to unfold (with the appearance of suddenness interpreted as an artifact of an imperfect fossil record), and they only made the ordinary occur faster (more intense Darwinian competition in tough times, for example, leading to even more efficient replacement of less adapted by superior forms). The reinterpretation of mass extinctions as central to life's pathway and radically different in effect began with the presentation of data indicating that the impact of a large extraterrestrial object set off the last great extinction at the Cretaceous- Tertiary boundary 65 million years ago.

Although the Alvarez hypothesis initially received very skeptical treatment from scientists (a proper approach to highly unconventional explanations), the case now seems virtually proved by discovery of the "smoking gun," a crater of appropriate size and age located off the Yucatan peninsula in Mexico. This reawakening of interest also inspired paleontologists to tabulate the data of mass extinction more rigorously. Some researchers established that multicellular animal life experienced five major and many minor mass extinctions during its 530 million- year history.

We have no clear evidence that any but the last of these events was triggered by catastrophic impact, but such careful study leads to the general conclusion that mass extinctions were more frequent, more rapid, more extensive in magnitude and more different in effect than paleontologists had previously realised. These four properties encompass the radical implications of mass extinction for understanding life's pathway as more contingent and chancy than predictable and directional.

Mass extinctions are not random in their impact on life. Some lineages succumb and others survive-as sensible outcomes based on presence or absence of evolved features. But especially if the triggering cause of extinction be sudden and catastrophic, the reasons for life or death may be random with respect to the original value of key features when first evolved in Darwinian struggles of normal times. This "different rules" model of mass extinction imparts a quirky and unpredictable character to life's pathway based on the evident claim that lineages cannot anticipate future contingencies of such magnitude and different operation.

To cite two examples from the impact- triggered Cretaceous-Tertiary extinction 65 million years ago: First, an important study published in 1986 noted that diatoms survived the extinction far better than other single-celled plankton (primarily coccoliths and radiolaria). This study found that many diatoms had evolved a strategy of dormancy by encystrnent, perhaps to survive through seasonal periods of unfavourable conditions (months of darkness in polar species as otherwise fatal to these photosynthesising cells; sporadic availability of silica needed to construct their skeletons). Other planktonic cells had not evolved any mechanisms for dormancy.

If the terminal Cretaceous impact produced a dust cloud that blocked light for several months or longer (one popular idea for a "killing scenario" in the extinction), then diatoms may have survived as a fortuitous result of dormancy mechanisms evolved for the entirely different function of weathering seasonal droughts in ordinary times. Diatoms are not superior to radiolaria or other plankton that succumbed in far greater numbers; they were simply fortunate to possess a favourable feature, evolved for other reasons, that fostered passage through the impact and its sequelae.

Second, we all know that dinosaurs perished in the end Cretaceous event and that mammals therefore rule the vertebrate world today. Most people assume that mammals prevailed in these tough times for some reason of general superiority over dinosaurs. But such a conclusion seems most unlikely. Mammals and dinosaurs had coexisted for 100 million years, and mammals had remained rat-sized or smaller, making no evolutionary "move" to oust dinosaurs. No good argument for mammalian prevalence by general superiority has ever been advanced, and fortuity seems far more likely.

As one plausible argument, mammals may have survived partly as a result of their small size (with much larger, and therefore extinction- resistant, populations as a consequence, and less ecological specialisation with more places to hide, so to speak). Small size may not have been a positive mammalian adaptation at all, but more a sign of inability ever to penetrate the dominant domain of dinosaurs. Great revolutions in the history of science have but one common, and ironic, feature: they knock human arrogance off one pedestal after another of our previous conviction about our own self-importance.

In three examples, Copernicus moved our home from centre to periphery, Darwin then relegated us to "descent from an animal world"; and, finally (in one of the least modest statements of intellectual history), Freud himself discovered the unconscious and exploded the myth of a fully rational mind. In this wise and crucial sense, the Darwinian revolution remains woefully incomplete because, even though thinking humanity accepts the fact of evolution, most of us are still unwilling to abandon the comforting view that evolution means (or at least embodies a central principle of) progress defined to render the appearance of something like human consciousness either virtually inevitable or at least predictable.

The pedestal is not smashed until we abandon progress or complexification as a central principle and come to entertain the strong possibility that H. sapiens is but a tiny, late-arising twig on life's enormously arborescent bush - a small bud that would almost surely not appear a second time if we could replant the bush from seed and let it grow again. Primates are visual animals, and the pictures we draw betray our deepest convictions and display our current conceptual limitations.

Artists have always painted the history of fossil life as a sequence from invertebrates, to fishes, to early terrestrial amphibians and reptiles, to dinosaurs, to mammals and, finally, to humans. There are no exceptions; all sequences painted since the inception of this genre in the 1850s follow the convention. No scene ever shows another invertebrate after fishes evolved but invertebrates did not go away or stop evolving! After terrestrial reptiles emerge, no subsequent scene ever shows a fish (later oceanic tableaux depict only such returning reptiles as ichthyosaurs and plesiosaurs). But fishes did not stop evolving after one small lineage managed to invade the land.

In fact, the major event in the evolution of fishes, the origin and rise to dominance of the teleosts, or modern bony fishes, occurred during the time of the dinosaurs and is therefore never shown at all in any of these sequences -even though teleosts include more than half of all species of vertebrates. Trees - or rather copiously and luxuriantly branching bushes-rather than ladders and sequences hold the key to this conceptual transition. We must learn to depict the full range of variation, not just our parochial perception of the tiny right tail of most complex creatures.

We must recognise that this tree may have contained a maximal number of branches near the beginning of multicellular life and that subsequent history is for the most part a process of elimination and lucky survivorship of a few, rather than continuous flowering, progress and expansion of a growing multitude. We must understand that little twigs are contingent nubbins, not predictable goals of the massive bush beneath. We must remember the greatest of all Biblical statements about wisdom: "She is a tree of life to them that lay hold upon her; and happy is every one that retaineth her."

Family Tree

The Family Tree that was expressed before permits the classification of all living beings to be a practically exact discipline. Most contemporary biologists assure that life has only emerged once on Earth. Every living being, if a virus, bacteria, mushroom, plant or animal, descends from a unique common ancestor. The evidence provided by the chemical characteristics of genetic material that determines the formation of every living being, points at, without a doubt, towards that common origin.

In each one of our cells we contain sets of genes, grouped in chromosomes, that repeat the instructions for constructing a living being an unaccountable amount of times, also detailing with care the characteristics of the biochemical functioning of all of that being. Many of those sets of instructions are as identical in bacteria as in a turtle of even a human being.

An organism will be more related with an other as they share more genetic material. Even though the evolution of species is produced thanks to the nucleic acids, it has been discovered that their exists a process of constant change in the characteristics of some of the chemical substances produced by the organism, which permits the dating of the time at which a certain species separated from an other or some common ancestor of various species that now constitute a family. It really results impacting to know that the approximate date, which is sometimes millions and even billions of years ago, in which their lived a common ancestor of us and a domestic animal, or the poor beast from which came the meat we are eating, or in between any vegetable or microscopic organism that we could think of. The British biologist Richard Dawkins points out some interesting dates about this.

The last common ancestor between men and chimpanzees lived as recently as five million years ago. The last common ancestor between all anthropoids - men, gorillas, and chimpanzees - and all the monkeys of the old and new continent (apes with a tail) lived some 35 million years ago. The last common ancestor between us and the lemurides lived 40 million years ago. Our last common ancestor with the carnivores (cats, dogs) lived some 60 million years ago. At about 65 million years ago their lived the last common ancestor between men and whales, bats, horses, cows, elephants, and other mammals of this type.

At least 100 million years ago their lived our last common ancestor with the insectivores (moles, hedgehogs). The last common ancestor between men and kangaroos live 150 million years ago, and 170 million years ago the last ascendancy of humans and platypuses. Many years before, in the carboniferous period, 320 million years ago, their lived the last common ancestor of man and all birds and reptiles. At about 370 million years ago, in the Devonian period, the last common ascendancy between humans and frogs died out, and 420 millions years ago the last one between the scaled fishes and ourselves disappeared. A "short" time before, 470 millions years ago, the last common ancestor of all vertebrates died.

About a billion years ago (one thousand million) the last common ancestors of chordates - ourselves - , arachnids, worms, insects, and, a little before, the last one's between insects and mollusks. The last common ancestor between a whale and a pea lived about 2 billion years ago, without a doubt, a unicellular organism similar to bacteria. The oldest common ancestor of all lived about 3 billion years ago.

But not all agree that life only originated once Earth and that all living beings constitute a single family which is sometimes not a happy one. There are indications that at least another type of life exists, which has an activity that has proven to be very harmful against beings based on the chemistry of DNA like us. According to the American pathologist Lewis Thomas, one of the most terrible sicknesses of moderns times is caused by microorganisms with a chemical base radically different to DNA.

More than two million people worldwide are affected by some senile sickness. Individuals that suffer sicknesses like Alzheimer's and Jacob or Creutzfeld, have suffered an invasion in their cerebral tissue by trillions of microorganisms similar to "slow viruses", which have no nucleic acids but that can, nevertheless, fabricate replicates of themselves. They are profoundly aggressive living beings, with whom we are not related at all and with whom we don't share any ancestor.

It is almost certain that all of the life forms that live on Earth today originated in the way described, but do we really understand the concept of the long period of time that has gone by ever since the origin of the universe and earthly life forms.

2

Biological Course

Pituitary tumours are common neo-plasms that exhibit a highly variable biological course as evidenced by hormonal and proliferative activities. Progress in defining the factors that govern cell differentiation in the pituitary have led to a new classification of anterior pituitary adenoma tumour cell types. Traditionally, there had been an on-going controversy regarding the basis of pituitary tumorigenesis. Two prevailing theories confronted hormonal stimulation against an intrinsic pituitary defect.

Several animal models and unusual clinical cases have provided support for the role of hormonal stimulation in the development of these neo-plasms and there is evidence for intra-pituitary production of hypothalamic hormones that may be responsible for excess stimulation. In contrast, the clonal nature of pituitary adenomas and the lack of associated hyperplasia accompanying pituitary adenomas strongly argue for an intrinsic somatic defect(s) as the principal etiology contributing to the genesis of these lesions.

Evidence suggestive of a possible hormonal etiology in

pituitary tumorigenesis includes paradoxical pituitary hormone responses to exogenous peptide stimulation, the development of pituitary adenomas in situations of excessive hypothalamic hormone stimulation or loss of negative feedback inhibition by target gland hormones, and evidence of hypothalamic hormone production within the anterior pituitary. Persuasive arguments against a hormonal etiology in human pituitary tumorigenesis includes the rarity of hyperplastic changes associated with pituitary adenomas, the lack of true adenomatous changes in the pituitary even after sustained exposure to hypothalamic hormones, and the relatively low frequency of recurrence following successful adenoma resection.

Growth Hormone Releasing Hormone

Growth hormone-releasing hormone (GHRH). GHRH can cause somatotroph proliferation and somatotroph hyperplasia is well documented as a consequence of chronic stimulation in patients with endocrine carcinomas ectopically secreting GHRH. Intra-pituitary GHRH expression is well documented and GHRH may be over-expressed in some aggressive tumours.

In vitro, human somatotroph adenomas respond to GHRH stimulation consistent with functional GHRH receptors on these tumours. In some instances, however, they may lose the down-regulation characteristic of normal somatotrophs. Thus, it would appear that GHRH stimulation may play a contributory role in the development of pituitary adenomas. Moreover, older transgenic mice over-expressing GHRH develop pituitary adenomas. However, in situations of GHRH excess, GH-producing adenomas are usually accompanied by somatotroph hyperplasia which is distinctly unusual in sporadic human pituitary GH adenomas.

Moreover, sustained exposure to ectopic GHRH does not always lead to true adenoma formation. Cloning of the GHRH receptor has permitted examination of the role of GHRH in somatotroph function. Indeed, loss of GHRH receptor signalling is now recognised as the genetic basis for the little (lit/lit) dwarf mouse. A truncated alternatively spliced form of the GHRH receptor with limited signalling properties has been identified in GH-producing pituitary adenomas. The GHRH receptor expression does not appear to be restricted to somatotroph-derived adenomas suggesting a potential non-GH specific role for this receptor in the pituitary.

Unlike the case with other examples of endocrine hyper-function where constitutive activation of the relevant receptor has been described, no intrinsic constitutively active forms of the GHRH receptor have thus far been identified in pituitary adenomas.

Corticotropin-Releasing Hormone (CRH) : The postulated etiology of Cushing's disease has shown tremendous flux since Cushing's original description of the disease. The documentation of adrenal hyper-responsiveness to ACTH and the presence of Crooke's hyalinisation in the pituitary brought a primary adrenal etiology to the fore. It has been recognised in the last few decades that patients with Cushing's disease may have other associated neuroendocrine abnormalities. Reports of therapeutic response to anti-serotoninergic or anti-dopaminergic agents reverted attention to the hypothalamus.

Long term follow-up of patients who have undergone trans-sphenoidal resection of microadenomas has indicated recurrence of disease in some patients. A few patients with pituitary Cushing's disease have corticotroph hyperplasia as the cause of the disorder in the absence of a discrete adenoma. These findings have implicated CRH excess in the pathogenesis of Cushing's disease. The characterisation of CRH in 1981 permitted its identification in a number of extra-pituitary tumours associated with a clinical picture resembling Cushing's disease; some of these patients had corticotroph hyperplasia.

In one instance, a hypothalamic gangliocytoma producing CRH was associated with corticotroph hyperplasia and Cushing's disease. Continuous infusion of CRH leads to corticotroph hyperplasia, but as of yet, has not been shown to result in true adenoma formation. CRH treatment of pituitary adenomas in vitro induces POMC and ACTH mRNA gene expression which are inhibited by dexamethasone. Additionally, CRH receptor expression appears not only intact in ACTH-producing pituitary adenomas but, unlike in the rat pituitary, is up-regulated in response to CRH treatment. There is currently no evidence of constitutive activation of CRH receptors in corticotroph adenomas. The closely related vasopressin V3 receptor is also intact but may be over-expressed in some corticotroph adenomas where it may play a role in tumour development.

Thyrotropin-Releasing Hormone (TRH) : Primary hypothyroidism is a well recognised cause of pituitary thyrotroph and lactotroph hyperplasia. These patients exhibit the spectrum of hyperplasia-to-neoplasia, suggesting that continuous stimulation by TRH may lead to thyrotroph adenoma. TRH has also been shown to be expressed in the pituitary and by the different types of pituitary adenomas. TRH signalling appears to be intact in pituitary adenomas as evidenced by intact binding and release of TSH and PRL. TRH receptor expression and structure is grossly unaltered even in thyrotroph adenomas.

However, messenger RNA is alternatively spliced in some pituitary tumours. Deletion of exon 3 results in a truncated product which does not bind TRH nor is it activated by it. The relatively higher levels of the truncated forms compared to the full length form of the TRH receptor in lactotroph adenomas may explain some of the paradoxical in vivo responses to TRH administration.

Gonadotrophin-Releasing Hormone (GnRH) : The occurrence of gonadotroph adenomas in patients with hypogonadism has suggested that the chronic stimulation resulting from primary gonadal failure may play a role in the formation and growth of these adenomas. Nevertheless, the majority of gonadotroph adenomas are not associated with underlying primary hypogonadism nor is there evidence of chronic hypothalamic stimulation in the adjacent non-tumorous adenohypophysis. Both GnRH and GnRH receptor expression have been documented in the different types of pituitary adenomas. Furthermore, pituitary adenomas with truncated GnRH receptors have been described that fail to respond to GnRH stimulation by enhancing calcium transport and gonadotrophin release in vitro. No activating mutations of the GnRH receptor, including the exon 3 "hot spot," have been identified.

Other releasing factors : The pathogenesis of lactotroph adenomas may involve defects in inhibitory hypothalamic factors or excessive stimulation by a putative PRL-releasing factor such as TRH or vasoactive intestinal peptide (VIP). The presence of lactotroph hyperplasia in the tissue surrounding lactotroph adenomas in some cases would be consistent with this theory. Moreover, lactotrophs normally proliferate during pregnancy thereby implicating estrogen as a candidate factor. Administration of oral contraceptives was implicated in the rapid increase in size and secretion of some lactotroph adenomas. The current cumulative clinical experience, however, has refuted this association with little evidence that low doses of oral contraceptives play a significant role in pituitary tumour development.

Loss of Inhibitory Hormones

Dopamine : The role of diminished hypothalamic inhibition was first suggested based on the observation of neo-vascularisation in lactotroph adenomas. It was speculated that neo-vascularisation would allow lactotrophs to escape from tonic dopaminergic inhibition. Dopamine signal transaction is mediated through D1 receptors which stimulate adenylyl cyclase activity and D2 receptors (D2R) which inhibit this enzyme. The family of dopamine receptors is much more complex in terms of biochemical, physiological, and pharmacological diversity.

Nevertheless, it appears that the predominant anterior pituitary dopamine receptor is the D2R. Activation of the D2R results in dysregulated cAMP production, potassium and calcium channel fluxes, phosphatidyl inositol turnover, and intracellular calcium concentrations. Treatment with EGF or TRH, which stimulates p44/42 MAP kinase, rescues GH3 cells from dopamine-induced apoptosis, with concomitant inhibition of dopamine-induced p38 MAP kinase activation. These findings are consistent with the view that dopamine induces apoptosis through p38 MAPK activation, and that the p44/42 MAP kinase signalling through growth factors has an opposing effect on p38 MAPK as well as on apoptosis.

Selective elimination of D2R activity in D2R knockout mice results in lactotroph hyperplasia and, subsequently, lactotroph adenoma formation in female D2R-deficient mice at 17 to 20 months of age. Interestingly, these lesions are monohormonal PRL immunoreactive neo-plasms that display characteristic juxtanuclear Golgi pattern of PRL staining and loss of the reticulin fibre network. Several of these adenomas have been noted to be much larger than normal glands with marked suprasellar extension and invasion of brain but no gross evidence of distant metastases.

In contrast, the formation of adenomas in male D2R-deficient mice without pre-existing or concomitant hyperplasia suggests that prolonged loss of dopamine inhibition may cause neo-plastic transformation by a distinct cellular mechanism than in female animals. While some pituitary tumours have been shown to be responsive to dopamine suppression, the dopaminergic resistance that is found in some of these tumours implicates diminished DR activity as a putative etiological factor in pituitary tumorigenesis. Thus far, however, investigation of the D2R gene has revealed it to be structurally intact in human lactotroph adenomas as well as in adenomas that secrete GH or TSH.

Somatostatin : GH secretion is under opposing influence from hypothalamic stimuli including GH-releasing hormone

(GHRH), which stimulates, and somatostatin (SS), which inhibits GH secretion. Specific receptors for SS (SSTRs) are expressed on somatotroph adenomas. Earlier studies suggested a relationship between the density of SS receptors on GH rumours and the secretory response to this analogue both in vitro and in vivo.

Binding sites for SS, however, have been identified by autoradiography in tumours resistant to the GH lowering effects of octreotide. These findings are consistent with differential adenylyl cyclase coupling by the five known subtypes of SSTRs and their heterogenous expression in pituitary adenomas. Expression of SS itself in large invasive GH tumours appears to be diminished compared to that in the normal pituitary. Taken together, these findings suggest multiple paracrine, autocrine, as well as endocrine mechanisms for SS-mediated control of somatotroph function and proliferation.

Glucocorticoid Hormones : As with other pituitary axes, primary adrenal insufficiency, with prolonged glucocorticoid deficit leads to corticotroph hyperplasia and rarely early adenoma formation. Arole for CRH in mediating this cell proliferation cannot be excluded. Lack of suppressibility of corticotroph adenomas by glucocorticoids had been suggested as a possible mechanism involved in the pathological ACTH secretion in Cushing's disease and Nelson's syndrome.

The human glucocorticoid receptor (GCR) pre-mRNA is alternatively spliced to generate a GCR a isoform and the N-terminally closely related *B* isoform. The isoform, however, differs in its 50 amino acid C-terminus which contains a unique 15-amino acid sequence that hinders glucocorticoid binding and gene transactivation. The functional inter-relationship between the two GCR isoforms in the pituitary and pituitary adenomas will undoubtedly be the focus of future studies.

Nevertheless, a molecular basis for glucocorticoid insensitivity has already been described in association with generalised or selective loss of function. Specific point mutations resulting in diminished ligand binding in the glucocorticoid hormone-binding domain are now known in cases of familial glucocorticoid resistance. Additionally, a novel germ line mutation has been reported to result in pituitary Cushing's disease. Similarly, rare reports of somatic mutations in the GCR with diminished glucocorticoid inhibition were noted in Nelson's syndrome and as predicted in cases of ectopic Cushing's syndrome.

Thyroid Hormones : The development of pituitary thyrotroph adenomas in patients with prolonged primary hypothyroidism has provided further evidence supporting the hypothalamic role in pituitary tumorigenesis. Thyroid hormones mediate their actions via nuclear thyroid hormone receptors (TRs) that bind to specific regulatory hormone response elements. There are two major classes of TRs, designated as α and β, each of which undergoes alternative splicing to generate $\alpha 1$ and $\alpha 2$ and $\beta 1$ and $\beta 2$ isoforms. With the exception of the p2 form, which is predominantly expressed in the hypothalamic-pituitary system, these receptor isoforms are ubiquitously expressed.

Of interest in the pituitary, the $\beta 1$ and $\beta 2$ isoforms appear to be expressed to a lesser extent in endocrinologically-inactive adenomas compared with the normal gland. Screening TRα mRNA identified three novel missense mutations, two in the common TRα region and another that was $\alpha 2$ specific. TRβ response elements failed to show any differences from published sequences. In contrast, mice with targeted disruption of the entire TR-β locus exhibit elevated thyroid hormone levels as a result of abnormal central regulation of thyrotropin but do not develop pituitary tumours. Thus, the putative differential hormone regulatory and mitogenic effects of the different THR isoforms in the pituitary remains largely understood. *Gonadal steroids:* The development of pituitary gonadotroph adenomas in patients with prolonged primary hypogonadism suggests that the lack of hormone negative feedback may facilitate pituitary tumour formation. Again, however, the role of GnRH stimulation cannot be easily distinguished from that of gonadal hormone inhibition in the development of these adenomas.

Growth Factors and Receptors : Growth factors are polypeptides of several major families that regulate cell replication and functional differentiation by directly altering the expression of specific genes. They are considered to play an important role in the multistep pathway of tumorigenesis. A number of oncogene products are homologous to growth factors, their receptors, or enzymes that participate in the mitogenic process. In several systems, growth factors have been shown to interact with specific membrane receptors in regulating cell growth and gene expression in an autocrine or paracrine manner. Some are known to affect hormone production and some are, in turn, modulated by hormones. A few have been identified in the hypothalamus and are considered to play a physiological role in pituitary regulation. The pituitary is a site of both synthesis and action of growth factors.

A number of growth factors have been identified in adenohypophysial cells, including insulin-like growth factors-I and -II (IGF-I, IGF-II), epidermal growth factor (EGF), nerve growth factor (NGF), transforming growth factor-α (TGF-α), transforming growth factor-β (TGF-β), and basic fibroblast growth factor (bFGF). Several partially characterised pituitary-derived growth factors have also been described, including thyroid hormone inducible growth factor, mammary cell growth factor, adrenal growth factor, chondrocyte growth factor, and adipocyte growth factor. Growing evidence suggests that human pituitary tumour cells produce multiple peptides which stimulate rat adenohypophysial cell replication in vitro.

The relative significance of these different growth factors in human pituitary adenomas remains to be established, however, several have been implicated in the pathogenesis of these tumours.

Epidermal Growth Factor : The EGF family of ligands includes EGF, TGF-a, amphiregulin, heparin-binding EGF-like growth factor (HB-EGF), and betacellulin (BTC). An additional family of EGF related agonists include neuregulins which include glial growth factors (GGFs), neu differentiation factors (NDFs)/heregulins, ligands for erbB-3 and erbB-4. Of interest, GGFs were purified from the bovine pituitary. It is currently not known, however, which specific subsets of erbB receptors become activated in response to each of these ligands.

Transforming growth factor-α : Transforming growth factor-α is expressed as a membrane-anchored protein by human adenohypophysial cells and tumours. TGF-α may alter pituitary production of GH, PRL, TSH as well as cell proliferation. Estrogen stimulation has been implicated in pituitary tumorigenesis and TGF-α appears to mediate some estrogenic effects. Targeted overexpression of TGF-α under the control of the PRL promoter results in lactotroph adenomas providing compelling evidence for the significance of this growth factor in pituitary tumorigenesis.

Epidermal growth factor and receptor (EGF; EGFR) : EGF is detectable by immunohistochemistry in most adenohypophysial cells and its mRNA is expressed with marked variation in all types of functional and non-functional adenomas. EGF potently stimulates PRL and ACTH secretion and has been reported to stimulate and inhibit GH secretion by non-tumorous rat pituitary cells in vitro. The selective expression and specific effects of EGF suggest that the pituitary is an important target site for this growth factor's action. The common receptor of EGF and TGF-α, EGF-R, is a 170-kD plasma membrane tyrosine kinase product of the protooncogene v-erbB. EGF-R is overexpressed in several types of human cancers and in most instances this overexpression is accompanied by TGF-α expression; expression of this receptor appear to correlate with tumour aggressiveness.

EGF-R is expressed by pituitary adenomas with the highest

levels detected in recurrent somatotroph adenomas and aggressive silent subtype 3 adenomas, suggesting a selective mechanism for the EGF/EGF-R family in the growth of aggressive somatotroph tumours. The importance of the EGF-R in the somatotroph has been further addressed recently in transgenic mice using expressing a dominant negative EGF-R mutant lacking the intracellular protein kinase domain (EGFR-tr). Directed EGFR-tr expression to GH- and PRL-producing cells resulted in dwarf mice with hypoplasia of these cell populations.

When EGF-R-tr over-expression was delayed to the post-natal period, however, no specific phenotype was observed. These findings point to EGF-R as an integral component in the differentiation of the somatotroph. The EGF-R is one of four highly homologous tyrosine kinase receptors which include erbB2/HER2/ neu/pl85, erbB-3 (HER3), and erbB-4 (HER4). Growing evidence in support of functional cross-talk between the different members of this receptor family is now well recognised. Ligand-induced stimulation can result in transphosphorylation of neu via EGF-R. Over-expression of a wild type EGF-R and heterocomplex formation with neu dramatically increases receptor autophosphorylation and binding of EGF.

Cytoplasmic positivity for neu can be identified in non-tumorous pituitary cells using an antibody to the intracytoplasmic domain of neu. No membrane staining is found using an antibody to the extracellular domain; the latter is said to reflect gene amplification. Neu mRNA expression has been described in the normal and tumorous pituitary. No differences in degrees of mRNA expression, however, have been noted between the different adenoma types and normal human pituitary tissue as examined by competitive PCR. As neu can be activated to an oncogene by a point mutation in the transmembrane region, nucleotide substitutions in this domain were investigated. Direct sequencing of codon 659 revealed no point mutations in any of the tumours.

Moreover, since amplification of neu has been noted in a number of human malignancies, DNA from these pituitary adenomas was examined by differential PCR. No detectable differences were noted between the neu gene and the single-copy reference gene INF-γ. These findings indicated that the neu gene is expressed in a homogenous pattern in pituitary cells and their adenomas but that this expression is not associated with gene amplification or activating mutations to suggest a direct role in pituitary tumorigenesis.

Transforming growth factor-β : Transforming growth factor (TGF)-β has been implicated in the regulation of normal and neo-plastic cell function. TGF-beta regulates the expression of various proteins, including p27Kipl (p27), a cell cycle inhibitory protein. TGF-beta 1/2/3 isoforms and the TGF-betareceptor are expressed in normal and adenomatous pituitaries. Dispersed pituitary adenomas cells show a biphasic response to TGF-β with changes in FSH secretion. The TGF-β family, however, is represented in at least three different forms in the pituitary.

Inhibins and activins consist of two homo- or heterodimeric polypeptide subunits derived from a common precursor; inhibin A (α - β A) and inhibin B (α - β B) selectively inhibit the release of FSH from pituitary gonadotroph cells whereas activin (β A- β B), activin A (β A β A) and activin B (β B- β B) stimulate its release. Inhibin subunits are expressed by pituitary gonadotroph adenomas and activin is known to stimulate hormone secretion by these tumours. Activin effects are mediated by two kinds of binding proteins, activin receptors and follistatin; the former are required for activin binding, but follistatin binds the protein resulting in decreased activity.

Activin receptors are expressed in gonadotroph adenomas and follistatin expression is reduced or absent in some suggesting the possibility of enhanced activin stimulation as a pathogenetic mechanism in the development of these common pituitary tumours. In contrast, activin treatment inhibited cell proliferation in some non-functioning primary human tumours with diminished follistatin mRNA expression. These findings suggest a functional but heterogenous role for the activin/ follistatin balance in modulating pituitary tumour cell replication.

Vascular Endothelial Growth Factor : Vascular endothelial growth factor (VEGF) also known as vascular permeability factor (VPG) was also one of the growth factors first isolated from the pituitary. VEGF exists in a number of isoforms in human and rodent tissues including VEGF206h/205r, VEGF189h/ 188r, VEGF165h/ 164r, VEGF145h/144r and VEGF121) that differ in their molecular masses and biological activities. The VEGF isoforms bind with two tyrosinekinase receptors, KDR/flk-1 and flt-1.

In addition, VEGF165 binds with co-receptor, neuropilin-1, which is expressed in human endothelial cells and several types of non-endothelial cells including solid tumours. Recent studies on the role of estrogen in the regulation of tumour angiogenesis demonstrated that this steroid induces neo-vascularisation in parallel with early induction of VEGF and the VEGFR2- (flk-1/KDR) protein expression in both blood vessels and non-endothelial cells. Moreover, estrogen-induced rat pituitary tumours in Fisher 344 rats express higher VEGF164 and neuropilin-1 levels compared to control untreated rat pituitaries. These findings suggest that over-expression of VEGF and its receptor (VEGFR-2) may play an important role in the early phases of estrogen induced tumour angiogenesis in the rat pituitary.

Fibroblast Growth Factors and Receptors

Fibroblast growth factors (FCFs) : Basic fibroblast growth factor (bFGF, also known as FGF-2) is one of an ever-expanding family of FGFs several of which possess mitogenic, angiogenic, and hormone regulatory functions. β FGF immunoreactivity was described originally in the non-hormone producing bovine pituitary folliculo-stellate cells; since β FGF has been shown to regulate GR, PRL and TSH secretion by the rodent pituitary, it was implicated in paracrine regulation within the pituitary. In the human pituitary, in contrast, β FGF is produced by adenohypophysial cells that comprise pituitary adenomas.

Pituitary-derived *B*FGF stimulates replication of PRL-secreting cells. Elevated circulating βFGF-like immunoreactivity is noted in patients with multiple endocrine neo-plasia (MEN)-l and in patients with sporadic pituitary adenomas. Moreover, in one mouse model, estrogen-induced pituitary tumorigenesis was associated with parallel increases in the expression of a pituitary tumour transforming gene (PTTG) as well as FGF-2.

Transgenic mice expressing FGF-2 under the control of the GH and the *a*-subunit promoters developed hyperplasia of several adenohypophysial cell types but not frank adenomatous changes. In Xenopus laevis oocytes, a 1.5 kb FGF-2 anti-sense (GFG) RNA complementary to the third exon and 3'UTR of FGF-2 mRNA has been implicated in FGF-2 mRNA regulation. The human homolog has been localised to the same chromosomal site as FGF-2 (chromosome 4, JO4513 adjacent to D4S430), confirming this as a human endogenous anti-sense gene. This GFG anti-sense gene also encodes a 35 Kd protein, which is highly homologous with the MutT family of anti-mutator NTPases. Pituitary tumours have been shown to express FGF-2 and GFG while the normal human pituitary expresses GFG but not FGF-2; GFG protein levels are higher in the normal gland than in most tumours. It shows gene structure and regions of complementarity between the mammalian sense FGF and anti-sense GFG genes are depicted.

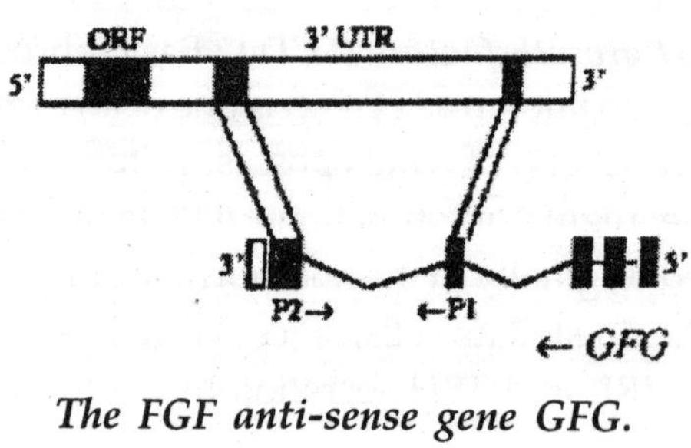

The FGF anti-sense gene GFG.

Aggressive pituitary adenomas appear to express more FGF-2 than GFG mRNA. Expression of GFG in transfected GH4 mammosomatotroph cells results in enhanced PRL gene expression and protein translation. Moreover, despite the fact that GFG expression does not down regulate pituitary FGF-2 mRNA expression, GFG expression inhibits pituitary cell proliferation. Taken together, these recent findings suggest that the GFG-encoded protein may represent a novel mechanism involved in restraining pituitary tumour cell growth while promoting hormonal activity.

Fibroblast growth factor receptors (FGFRs) : There are 4 mammalian FGFR genes encoding a complex family of transmembrane receptor tyrosine kinases (RTKs). Each prototypic receptor is composed of 3 immunoglobulin (Ig)-like extracellular domains, 2 of which are involved in ligand binding, a single transmembrane domain, a split tyrosine kinase, and a COOH-terminal tail with multiple autophosphorylation sites. Multiple forms of cell-bound or secreted receptors are produced by the same gene. Tissue-specific alternative splicing, variable polyadenylation sites and alternative initiation of translation result in truncated receptor forms. The first two extracellular loops of FGFR1 can be secreted as soluble circulating FGF binding proteins, but their physiological importance remains to be established. Different FGFRs can dimerise, so that truncated forms of FGFR1 block signalling through FGFR1, 2, and 3.

Structural alterations of FGFRs may play a role in human tumorigenesis. For example, FGFR1 is highly expressed in the brain but the shorter (2 Ig-domain) form of FGFR1 is more abundant in some CNS glioblastomas. Anti-sense targeted interruption of FGFR1 reduces malignant melanoma cell proliferation and differentiation. FGFR2 exon switching has been observed to accompany prostate cell transformation. The normal pituitary expresses mRNAs for FGFR 1, 2 and 3. An interesting finding was the documentation of novel truncated mRNAs for the first and second Ig-like loops of FGFR4 in the non-tumorous pituitary and a kinase-containing variant of

FGFR4 with an alternative transcription initiation site in pituitary adenomas. This tumour-derived kinase containing FGFR4 isoform is transforming in vitro and in vivo and has been recently shown to result in lactotroph adenomas in transgenic mice. Taken together, these data suggest that dysregulated FGF/FGFR function plays a role in pituitary : tumorigenesis and that FGFR4 is a candidate tumour-specific kinase.

Nerve Growth Factor Family : NGF overexpression targeted to lactotrophs results in dramatic hyperplasia of those cells, however, tumour formation has not been demonstrated. Further, treatment of human prolactinoma cells with NGF results in decreased proliferation in vitro, reduced capacity to form colonies in soft agar, and loss of tumorigenic activity in nude mice. NGF appears to induce D2 receptor expression in human prolactinomasand directs differentiation of bihormonal GH3 cells into mature lactotrophs with D2 receptor expression. Moreover, gp l40trk and gp75 components of the NGF receptor are expressed in responder prolactinoma cell lines. NGF antisense treatment results in loss of expression of D2 receptors and an increase cell proliferation. Aberrant expression of NGF may, therefore, contribute to unrestrained cell proliferation and/or diminished responsiveness to dopamine agonist treatment.

3

Different Forms of Life

Living Organisms

Classification, in biology, the identification, naming, and grouping of organisms into a formal system. The vast numbers of living forms must be named and arranged in an orderly manner so that biologists all over the world can be sure they know the exact organism that is being examined and discussed.

Groups of organisms must be defined by the selection of important characteristics, or shared traits, that make the members of each group similar to one another and unlike members of other groups. Modern classification schemes also attempt to place groups into categories that will reflect an understanding of the evolutionary processes underlying the similarities and differences among organisms.

Such categories form a kind of pyramid, or hierarchy, in which the different levels should represent the different degrees of evolutionary relationship. The hierarchy extends upwards from several million species, each made up of individual organisms that are closely related, to a few kingdoms, each containing large assemblages of organisms, many of which are only distantly related.

To construct classification schemes that correspond as closely as possible to the natural world, biologists examine and compare the anatomy, functions, genetic systems, behaviour, ecology, and fossil histories of as many organisms as possible. More than 1.5 million different groups have been identified and at least partly described, and many more remain to be studied.

All branches of biology contribute to such studies; the specialities that are immediately concerned with the problems of classification are taxonomy and systematics. Although the two disciplines overlap considerably, taxonomy is more involved with nomenclature (naming) and with constructing hierarchical systems, and systematics with uncovering evolutionary relationships. The systematic classification of living, or once living, things - which these days is done to be consistent with their presumed evolutionary relationships ("phylogeny") - is called taxonomy.

It is obvious that there exists a hierarchy of phylogenetic relationships among living things. Our very bodies exemplify such a hierarchy. "Naturalists have striven perpetually to build a meaningful classification scheme for living things. Long before Darwin, plants and animals were believed to be the primary divisions of life. In 1866, Haeckel was the first to challenge this dichotomy by suggesting that the Protista should be considered to be a third kingdom equal in stature to the Plantae and Animalia.

The Bacteria or Monera were designated as a fourth kingdom by Copeland, in 1938. Whittaker added the fungi in 1959, and his five kingdom (Plantae, Animalia, Protista, Fungi and Monera) universal tree is still taught as part of basic biology curricula. Over 50 years ago, Chatton, and Stanier and van Niel suggested that life could be subdivided into two even more fundamental cellular categories, prokaryotes and eukaryotes.

The distinction between the two groups was subsequently refined as studies of cellular biology and genetics progressed such that prokaryotes became universally distinguishable from eukaryotes on the basis of missing internal membranes (such as the nuclear membrane and endoplasmic recticulum), nuclear division by fission rather than mitosis and the presence of a cell wall.

The definition of eukaryotes was broadened to include Margulis' endosymbiont hypothesis, which describes how eukaryotes improved their metabolic capacity by engulfing certain prokaryotes and converting them into intracellular organelles, principally mitochondria and chloroplasts. By digesting in vivo labelled 16S rRNA using T1 ribonuclease then accumulating and comparing catalogues of the resultant oligonucleotide 'words', Woese and Fox were able to derive dendrograms showing the relationships between different bacterial species.

Analyses involving some unusual methanogenic 'bacteria' revealed surprising and unique species clusterings among prokaryotes. So deep was the split in the prokaryotes that Woese and Fox proposed in 1977 to call the methanogens and their relatives 'archaebacteria', a name which reflected their distinctness from the true bacteria or 'eubacteria' as well as contemporary pre-conceptions that these organisms might have thrived in the environmental conditions of a younger Earth.

In 1990, Woese, Kandler and Wheelis formally proposed the replacement of the bipartite view of life with a new tripartite scheme based on three kingdoms or domains; the Bacteria (formerly eubacteria), Archaea (formerly archaebacteria) and Eukarya (formerly eukaryotes although this term is still more often used). As humans, we have uniquely large brains and upright posture; as primates, fingernails and stereoscopic vision; as mammals, warm blood and milk-fed young; as amniotes (the group that includes mammals, birds, and reptiles) internal fertilization and the ability to reproduce ourselves outside a watery environment.

All animals have tissues, and all of us collect our carbon and energy from organic compounds derived from all those plants, algae, and bacteria that can use either chemical energy or light to manufacture their own supplies out of non-living raw materials.

Hierarchical Levels

Biologists classify individual organisms at the basic level of the species, which is the only category that can be regarded as occurring in nature. The higher categories are abstract groupings of species. A species is composed of organisms that resemble one another in many important characteristics. Moreover, in organisms that have sexual reproduction, a species is made up of interbreeding populations that, ideally, cannot produce fertile offspring with members of any other species.

Species that do not interbreed with each other but are clearly related by important shared traits are grouped into a genus (plural, genera), and the separate species are given a two-word name (binomial nomenclature). The first word is the genus name and the second word is an adjective, usually descriptive or geographic. This means of naming was established in 1758 by the Swedish naturalist Carolus Linnaeus, the founder of modern taxonomy. He used Latin names because the scholars of his day communicated in that language.

Linnaeus gave humans the genus name Homo (man) and the species name Homo sapiens (wise man). To construct the hierarchy of classification, one or more genera are grouped into a family, families are grouped into orders, orders into classes, classes into phyla, and phyla into kingdoms. The groups of organisms within these seven major categories, at every level of the hierarchy, are termed taxa, and each taxon has a definition that encompasses the important traits shared by all its constituent taxa.

To allow further subdivision, the prefixes sub- and super- may be added to any category. In addition, special intermediate categories-such as branch (between kingdom and phylum), cohort (between class and order), and tribe (between family and genus)-may be used in complex classifications. On any level, a taxon should indicate a common evolutionary background, that is, all its members should have evolved from a common ancestor.

The taxon is then said to be monophyletic. Where an established taxon includes two or more members that have converged so that they have traits in common but have evolved from different ancestral lines, the taxon is said to be polyphyletic. An attempt is then usually made to divide and redefine the taxon so that monophyletic taxa result.

Taxonomic Ranks

We can distinguish and name several kinds of dogs or fruit trees, but probably not so many weevils or wildflowers. Our own large size inclines us to notice and recognise other large vertebrates. We are more likely to be familiar with elephants (or polar bears or kangaroos) which come to our zoos from half a world away, than with the insects in our own gardens. Anything too small to see is a 'microbe' - few of us can distinguish among them; few have much interest. Yet the diversity among 'microbes' far exceeds that of all larger organisms. The most fundamental division of living things is into three categories called superkingdoms; two are exclusively microbial, the familiar plant and animal kingdoms comprise only a small part of the third. The division of life into plants and animals is a rough human classification that works well enough to describe various shapes and expected behaviours but has little to do with how life arose, or how early organisms made a living.

Likewise, among living organisms, especially the small ones, the labels "plant" and "animal" represent two collections of features drawn from a much larger total range, and often found in combinations that ignore these simple categories.

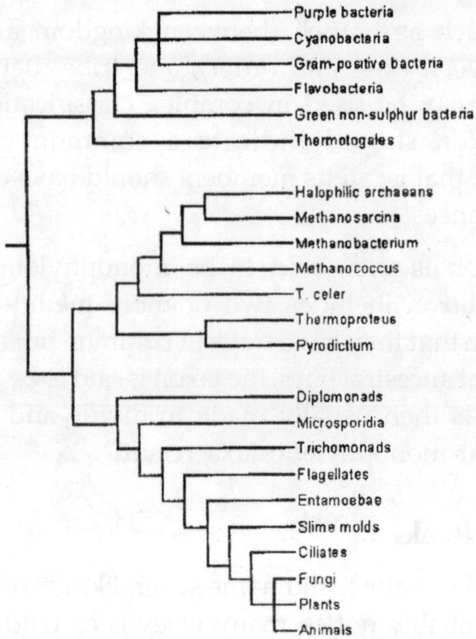

A rRNA (16S) cladogram showing presumed phylogenetic relationships between the three super-kingdoms: (a) Bacteria, (b) Archaea and (c) Eucarya (eukaryotes).

There are far more lifestyles than these two labels allow for, and some of them are practised by organisms that live off what we would consider to be inorganic foods. "Comparison of base or amino acid sequences in non-conserved portions of nucleic acids and proteins, especially the small subunit RNA molecule of the ribosome, have shown that life can be divided into three principal domains: the archaebacteria, eubacteria, and eukaryotes.

Some trees indicate a specific sister-group relationship between eukaryotes and archaebacteria. Regardless of the true branching topology of the primary domains, there can be little doubt that these three groups diverged from one another during the Archean Eon". First are the archaebacteria, "ancient bacteria." Not many species now survive.

They evolved on an alien Earth, and now live in extreme

environments: hot springs where temperatures never fall below 55°C; salt flats where salinities are four times ocean levels; the intestines of cattle or our own, for that matter.

Second are the "eubacteria," many thousands of species with an enormously broad variety of lifestyles fermenters, nitrogenfixers, sulphur users, oxygen producers, recycling specialists for all kinds of vital molecules. Both archaebacteria and eubacteria are prokaryotes, "preseeds," so called because of the way they carry their DNA loose inside their outer cell membrane.

The third fundamental division of life is the "true seeds" (eukaryotes, EKs for short), which package their genes in a "seed" or nucleus, a separate envelope inside the outer membrane. They comprise all of what we consider to be "higher organisms," including ourselves. There is one other problematic group, the viruses, which are particularly hard to define. They do not have cells, and probably stem from bits of DNA or RNA that somehow escaped from one or another of the three basic groups.

Viruses certainly do not grow, and they feed only if we greatly expand our definition of feeding. In fact, they exist only because they can replicate, and the materials they use in order to replicate are what causes problems; these refugee bits of genetic material' when bound with proteins, can replicate themselves only by reinvading and subverting living cells. Classification - Associating organisms together into various groupings which reflect, as best we know, the relationships between them.

There is a hierarchy of such groups, in which the most general group for living things is called a "superkingdom" and the most specific grouping called, of course, a "species." Actually, there can be several finer, less formal, groupings within a species - subspecies, variety, and so on - but for our present purposes it is sufficient to recognise only the following hierarchical divisions:

Most general; least specific	Super kingdom	There are three super kingdoms: Bacteria, Archaea and Eucarya.
	Kingdom	There are five or so kingdoms, namely animal, plant, fungi, Protoctista, and procaryotes.
	Phylum	The animal phylum Chordata includes all vertebrates. Two other examples are Arthropoda (insects, crustaceans, etc.) and Pteridophyta (ferns & related plants).
	Class	Vertebrata (a class within the Chordata) are animals with backbones. The trilobites (class Trilobita) are a subdivision of the Arthropoda.
	Order	The Carnivora includes cats, dogs, bears, etc., and is a subdivision of the Vertebrata. ...
	Family	
	Genus	
Least general;	most specific	Species

Phylogenetic Meaning of Hierarchy

In general, the higher taxonomic levels reflect more ancient ("deeper") relationships in the evolution of organisms. Employing the useful metaphor of a family tree, it is intuitively reasonable to suppose that the cat and dog family branches diverged from one another before the different kinds of cats (or dogs) began to separate as twigs on their respective branches.

Divergence precedes diversification : However, whereas the hierarchy may indicate the sequence in which various groups became differentiated, it tells nothing about the timing. Lions and tigers became distinct from one another after the cat/dog schism; we cannot say how long ago either event took place and, in particular, we cannot say that lions and tigers separated at the same time as dingos and wolves.

There are further complications also. Even supposing that we have a perfect understanding of how organisms evolved, which of course we do not, it must be remembered that taxonomic names are primarily defined by morphological criteria rather than phylogeny. Thus, although birds and mammals arose as subgroups within the reptiles, they are both accorded the status of 'class' - the same taxonomic rank as reptiles themselves - rather than some subordinate rank within the Reptilia.

If, as is generally believed, all life is descended from a single (or very few) common ancestor, such a compromise is necessary to avoid an unmanageably large number of taxonomic levels. If we had understood modern evolutionary synthesis during Linnaeus' time, it is possible we would not have adopted the "binomial" convention; we would have been influenced by the knowledge that phylogenetic groupings occur at variable levels and no two are strictly comparable.

Thus, the point where one draws the line between a generic distinction and a familial distinction is more or less arbitrary and is unlikely to be consistent between different lineages: Man classifies, Nature does not. Only the bio-species concept has an objective definition - can these two organisms produce viable offspring - and even that is by no means the black-and-white distinction one might think.

Two further criticisms may be levelled in respect to the Linnean hierarchy. The first criticism is that its use takes for granted a branched topology of relationships. This may be true for very large segments of the phylogenetic history of living beings, but it is definitely not true in several instances. First, the very origin of the eukaryotic cell, hence an event at the root of a disproportionately major branch of the phylogenetic tree, is currently explained as a symbiotic event, that is, as an event determining a cluster of anastomoses among the oldest branches of the tree of life.

Second, anastomoses of branches of the phylogenetic tree are produced by any successful event of hybridisation, which is possibly rare in animals but is certainly common in plants (at least in some groups). In all these instances, reducing the real topology of phylogeny to the conventional branched topology of the Linnean hierarchy is hardly 'natural'. The second criticism of the Linnean hierarchy comes from cladistics in particular. The problem is that a phylogenetic reconstruction may only allow for the identification of nesting relationships, but cannot offer any ground to the recognition of absolute ranks. For instance, the brown bear (Ursus arctos) will turn out to be a terminal twig of the bear family (Ursids), this being in turn a branch of the Carnivores, which are nested into a larger branch Mammals, and so on. However, nothing justifies giving the same rank (say, ordinal) to Carnivores and Rodents, or- outside Mammals - to Galliforms and Coleoptera.

What matters more is the inadequacy of the formal Linnean hierarchy to express the relationships of the clades in the system. This is currently a matter of dispute within the systematic community, especially with regard to what remains of the third school, known as the school of evolutionary systematics, whose principles are traditionally identified with those of E. Mayr and George Gaylord Simpson. According to this school, classification must be rooted in evolution, but the relevant evolutionary information must not be limited to genealogy but also extended to adaptation.

Accordingly, symplesiomorphies are not discarded as non-informative and paraphyletic groups are retained. As practical examples of the contrast between the two schools, two well-known groups that are clearly, or quite probably, paraphyletic and should therefore be discarded, or redefined according to cladistic criteria, are Reptiles and Dicotyledons. All of that notwithstanding, we are stuck with the Linnean naming convention which, for all its faults, succeeds in providing a useful shorthand for identifying taxa.

Morphology and Genomics

While comparative molecular data have helped, they have not proved as unambiguous as we had hoped. Molecules, like morphologies, vary in their evolutionary rates and are subject to parallel and convergent evolution: and in consequence different molecules often suggest different phylogenies, just as do different morphological characters. The most commonly used molecule is 18S rRNA, comprising about 1800 base pairs, because it is a slowly evolving molecule.

Slow evolution is a prerequisite for probing very ancient phylogenetic events, to avoid the saturation problem. However, this same property makes the molecule unsuited for distinguishing events which occurred close together, perhaps during a rapid radiation. A possible example of this problem is found among annelids and molluscs; analyses which include many representatives of both phyla show a complete mix of the two groups. (Echinoderms, conversely, always appear as a monophyletic group, suggesting the modern forms represent a single lineage which diverged long ago, accumulating many unique mutations.)

The historical development of taxonomic concepts did not proceed all at once, or even linearly. But not higher ranks in the mid-1700s; and yet the three superkingdoms -the broadest division of living things - were not recognised until the late twentieth century.

Recognising and naming species, however, does not exhaust the performance of folk taxonomies. Explicitly or not, these also include an element of hierarchy. That is, the basic named units are grouped into more general kinds, e.g. winged animals, scaly animals, etc., or legumes, lilies, etc. These 'genera' may be grouped, in turn, into still more general kinds, e.g. land animals, water animals, or herbs, trees, mushrooms. Up to five hierarchically nested ranks can be recognised in the most developed folk taxonomies.

The first half of the eighteenth century was dominated by the classification of plants offered in his Institutiones rei herbariae by Joseph Pitton the Tournefort, who arranged 698 genera in 122 sections in 22 classes. This classification was later overturned by Carolus Linnaeus, whose major treatises were

Species plantarum and Systema naturae. Both works went through many editions, but the first edition of Species plantarum and the tenth edition of Systema naturae were eventually adopted in the nineteenth century as the starting points of modern botanical and zoological nomenclature, respectively.

The success of Linnaeus did not simply derive from the sheer number of species he dealt with, nor from his work being the last to provide a single-author comprehensive survey of nature, but primarily from his straightforward use of a clear and uniform hierarchical arrangement of species and groups and his systematic adoption of binomial nomenclature. A critical contribution towards the future developments of biological systematics was provided by Georges-Louis Leclercq, comte de Buffon, who suggested a biological, i.e. functional, definition of species.

In his concept, a species is a reproductive community: members of this community may freely interbreed, thus generating fertile offspring, whereas members of different species, even if similar, cannot interbreed or, if they can, only generate sterile hybrids, such as the mules and hinnies obtained by crosses between horse and donkey. During this time there was much debate about the possibility of developing a natural, i.e. non-arbitrary, system of living beings. This possibility gained support with the development of evolutionary thinking. The otherwise mysterious 'affinities' between species could then be explained by common descent.

In due course the reconstruction of phylogeny (genealogical relationships) become the necessary background for any serious systematic endeavour. The phylogenetic (or cladistic) school largely identified itself, at the beginning, with Hennig's manifesto Phylogenetic Systematics. With Willi Hennig, the reconstruction of phylogeny becomes the primary aim of the systematist. Phylogenetic relationships are expressed by tree-like representations called cladograms.

To reconstruct phylogeny, only shared derived features

(synapomorphies) are informative, whereas shared primitive features (symplesiomorphies) are not. Synapomorphies identify natural units (monophyletic taxa or clades), whereas symplesiomorphies identify 'incomplete' groups (paraphyletic taxa or grades). Methods have been developed, since Hennig, both for the identification of the polarity of character states (primitive or plesiomorphic versus derived or apomorphic) and for the reconstruction of phylogeny from a matrix of polarised character states. Only monophyletic taxa are accepted by cladists. Strictly speaking, what is obtained by the hierarchical arrangement of nested clades should be called a system, rather than a classification.

Rules of Taxonomy

The rules for formal nomenclature require designation of types, which permanently delimit, by their measurable properties, that which must be included within the boundaries of the taxon; descriptions, which, unlike the types, can be altered (= emended); and, for fossils, illustrations.

Today, biological nomenclature is governed by the following codes:

- International Code of Zoological Nomenclature (ICZN)
- International Code of Botanical Nomenclature (ICBN)
- International Code of Nomenclature of Bacteria (ICNB).

There are also an International Code of Nomenclature for Cultivated Plants and a Classification and Nomenclature of Viruses, as Fifth Report of the International Committee on Taxonomy of Viruses. Several times, beginning with the 1840s and finishing with the 1990s, there have been efforts towards devising a single code which would rule the nomenclature of all living beings, but no workable result has been yet achieved.

"It is occasionally suggested, especially because of problems with protists, that there should be just one nomenclatural code, and that would have seemed rational to Linnaeus. The differences between the two major codes, the ICBN and ICZN, are not so great as to make a unification impossible, but the prospect is unlikely. Indeed, there always have been pressures to multiply codes, resulting already, for example, in a separate code for bacteria".

Level of Subdivision

In the biological classification system, each level of subdivision has its own name. Here are the names for those levels for both the plant and animal kingdoms:

For Plants	*For Animals*
kingdom	kingdom
division	phylum
class	class
order	order
family	family
genus	genus
species	species

Note that the two breakdowns are identical, except that with plants the subdivision after kingdom is known as a "division," while the same level for animals is the phylum. Specialists refer to yet other subdivisions not shown, such as subfamilies, superorders, and subspecies or varieties, but these seven subdivisions are enough to make us happy here.

To give you a feeling for how these groupings "work," here's a simplified analysis of how they apply to the animal called the human being:

Kingdom : Animal

Phylu : Chordate (animals with backbones)

Class : Mammal (with hair, female makes milk)

Order : Primate (apes and monkeys)

Family : Mankind, (with extinct Neanderthals, etc.)

Genus : Homo

Species : Sapiens

You see, at every step down the classification ladder, the thing that we are is narrowed down. At first we're just animals. Then the phylum grouping separates us from all animals without backbones, such as sponges, insects, and worms. On down the ladder we go until we land at the species, and at that point we know that we're just talking about one kind of animal, and that animal is us.

Now let's do the same thing with House Sparrows:

Kingdom	: Animal
Phylum	: Chordate (animals with backbones)
Class	: Bird (egg-layers with feathers)
Order	: Passerine (songbirds)
Family	: Thick-beaked birds such as finches
Genus	: Passer
Species	: Domesticus

By comparing the two breakdowns, we can see that humans and birds are alike down to class level. We are both animals with backbones, but we humans sure don't belong to the animal class that lays eggs and has feathers. So, now it's clear that plant and animal names and plant and animal classification elegantly blend into one another.

The classification system also theoretically reflects the manner by which the various groups and species of organisms arose during the course of evolution. If we're talking about birds called shrikes and we say that the whole genus has hooked bills, we're implying something rather profound. We're suggesting that at a certain point in bird evolution a species arose with a hooked bill and other characteristics today associated with shrikes, and this species became the ancestor of all the many shrikes that exist today, the proto-shrike, the first of its genus.

Well, all that is pretty theoretical, so let's get back to the basics. Let's ramble a little about these various classification categories, just for practice: For instance we can say that the above breakdown shows that all birds belong to the same class. By the same token, all mammals, from mice to humans, belong to the same class. Among plants, all flowering plants comprise the angiosperm class, and all conifers, such as pines and spruces, comprise the conifer class. In the bird world, there are only around thirty orders; one order holds the woodpeckers, another the penguins, and another all the passerines, or perching birds.

In the plant world, only specialists deal with orders. Much more important are plant families, such as the oak family, the grass family, and the sunflower, or composite, family. Specialists seldom speak of "the shrew family" or "the moss division of plants." They use technical, Latinised words. The shrew family is the Soricidae, while the moss division of plants is known as the Bryophyta. Since we're beginners, however, we don't really need to wrestle with these terms. By the way, the science dealing with scientific names is called taxonomy.

Five Kingdoms

Two kingdoms of living forms, Plantae and Animalia, have been recognised since Aristotle, in the 4th century BC, established the first taxonomy. In their way of life and evolutionary path, rooted plants are so distinct from mobile, food-ingesting animals that the concept of the two kingdoms remained intact until recently.

Only in the 19th century, long after it was revealed that one-celled organisms could not fit comfortably into either of the two categories, was it proposed that unicellular forms be placed in a third kingdom, the Protoctista. Furthermore, long after photosynthesis was discovered to be the basic nutritional mode of plants, the fungi, which feed by absorption, continued to be classified as plants because of their apparently rooted manner of growth.

Recently, as techniques for examining the cell have improved dramatically, it has become clear that the major division in the living world is not that between plants and animals but between organisms whose cells have no enclosed nucleus and organisms whose cells have nuclei bound by membranes. The former are called prokaryotes ("before kernels") and the latter eukaryotes ("true kernels"). Prokaryotic cells are also lacking in organelles-mitochondria, chloroplasts, advanced flagella, and other special cell structures-at least some of which occur in all eukaryotic cells.

The bacteria and blue-green algae are prokaryotic cells, and they have been recognised in modern taxonomies as a fourth kingdom, Prokaryota, formerly known as the kingdom Monera. Eukaryotic cells arose much later and may have evolved as symbiotic associations of prokaryotes. The kingdom Protoctista is composed of diverse one-celled organisms, either free-living or colony-forming. Each of the multicellular kingdoms is believed to have arisen more than once from protoctist ancestors.

The kingdom Animalia comprises organisms that are multicellular, have their cells organised into different tissues, are mobile or partly mobile by means of contractile tissues, and digest food internally. The kingdom Plantae is made up of multicellular organisms that usually have walled cells and that contain chloroplasts in which they produce their own food by photosynthesis.

However, in 1999 preliminary results of a major cladistics study indicated that the plant kingdom should actually be re-classified as three kingdoms for green, red, and brown plants, as the study proposed that these three groups evolved from three different varieties of one-celled, plant-like organisms. The fifth kingdom, Fungi, comprises multicellular or multi-nucleate organisms that digest food externally and absorb it through the surfaces of protoplasmic tubes called hyphae.

This five-kingdom classification of the living world is thus based on three levels of organisation: the primitive prokaryotic; the relatively simple and primarily unicellular eukaryotic; and the complex, multicellular eukaryotic. Within the last of these levels, the three major directions of evolution are each based on a different kind of nutrition and are expressed in the different kinds of tissue organisation that are characteristic of animals, plants, and fungi.

Amazing Diversity of Living Systems

Living organisms are subdivided into 5 major kingdoms, including the Monera, the Protista (Protoctista), the Fungi, the Plantae, and the Animalia. Each kingdom is further subdivided into separate phyla or divisions. Generally "animals" are subdivided into phyla, while "plants" are subdivided into divisions. These subdivisions are analogous to subdirectories or folders on your hard drive. The basic characteristics of each kingdom and approximate number of species are summarised in the following table:

Prokaryotic Cells without Nuclei and Membrane-Bound Organelles

- ***Kingdom Monera [10,000 species]:*** Unicellular and colonial-including the true bacteria (eubacteria) and cyanobacteria (blue-green algae).
  Eukaryotic Cells With Nuclei And Membrane-Bound Organelles.
- ***Kingdom Protista (Protoctista) [250,000 species]:*** Unicellular protozoans and unicellular & multicellular (macroscopic) algae with 9 + 2 cilia and flagella (called undulipodia).
- ***Kingdom Fungi [100,000 species]:*** Haploid and dikaryotic (binucleate) cells, multicellular, generally heterotrophic, without cilia and eukaryotic (9 + 2) flagella (undulipodia).
- ***Kingdom Plantae [250,000 species]:*** Haplo-diploid life cycles, mostly autotrophic, retaining embryo within female sex organ on parent plant.
- ***Kingdom Animalia [1,000,000 species]:*** Multicellular

animals, without cell walls and without photosynthetic pigments, forming diploid blastula.

The five-kingdom system of classification for living organisms, including the prokaryotic Monera and the eukaryotic Protista, Fungi, Plantae and Animalia is complicated by the discovery of archaebacteria. The prokaryotic Monera include three major divisions: The regular bacteria or eubacteria; the cyanobacteria (also called blue-green algae); and the archaebacteria.

Lipids of archaebacterial cell membranes differ considerably from those of both prokaryotic and eukaryotic cells, as do the composition of their cell walls and the sequence of their ribosomal RNA subunits. In addition, recent studies have shown that archaebacterial RNA polymerases resemble the eukaryotic enzymes, not the eubacterial RNA polymerase.

Archaebacteria also have introns in some genes, an advanced eukaryotic characteristic that was previously unknown among prokaryotes. In eukaryotic cells, the initial messenger RNA (M-RNA) transcribed from the DNA (gene) is modified (shortened) before it leaves the nucleus. Sections of the M-RNA strand called introns are removed, and the remaining portions called exons are spliced together to form a shortened (edited) strand of mature M-RNA that leaves the nucleus and travels to the ribosome for translation into protein.

This process is known as "gene editing." Some authorities hypothesise that eukaryotic organisms may have evolved from ancient archaebacteria (archae = ancient) rather than from the common and cosmopolitan eubacteria. The archaebacteria could have flourished more than 3 billion years ago under conditions previously thought to be uninhabitable to all known life forms. Although many conservative references place the archaebacteria in a separate division within the kingdom Monera, most authorities now recognise them as a 6th kingdom—The kingdom Archaebacteria.

In fact, data from DNA and RNA comparisons indicate

that archaebacteria are so different that they should not even be classified with bacteria. Systematists have devised a classification level higher than a kingdom, called a domain or "super-kingdom," to accommodate the archaebacteria. These remarkable organisms are now placed in the domain Archaea. Other prokaryotes, including eubacteria and cyanobacteria, are placed in the domain Bacteria. All the kingdoms of eukaryotes, including Protista (Protoctista), Fungi, Plantae and Animalia, are placed in the domain Eukarya.

The large molecular differences between the majority of prokaryotes in the kingdom Monera and the archaebacteria warrants a separation based on categories above the level of kingdom. In other words, the differences between the true bacteria and archaebacteria are more significant than the differences between kingdoms of eukaryotes. The 3-domain system of classification is shown in the following table:

Three Domains (Super-kingdoms) of Living Organisms

- *Bacteria:* Most of the Known Prokaryotes
  - — *Kingdom (s):* Not Available at This Time
  - — *Division (Phylum) Proteobacteria:* N-Fixing Bacteria
  - — *Division (Phylum) Cyanobacteria:* Blue-Green Bacteria
  - — *Division (Phylum) Eubacteria:* True Gram Positive Bacteria
  - — *Division (Phylum) Spirochetes:* Spiral Bacteria
  - — *Division (Phylum) Chlamydiae:* Intracellular Parasites
- *Archaea:* Prokaryotes of Extreme Environments
  - — *Kingdom Crenarchaeota:* Thermophiles
  - — *Kingdom Euryarchaeota:* Methanogens & Halophiles
  - — *Kingdom Korarchaeota:* Some Hot Springs Microbes
- *Eukarya:* Eukaryotic Cells
  - — Kingdom Protista (Protoctista)
  - — Kingdom Fungi

— Kingdom Plantae
— Kingdom Animalia

The kingdom Protista includes a diverse array of organisms, from minute flagellated cells to macroscopic kelp. The smallest microscopic organisms are termed protists, consequently some biologists prefer to call this kingdom the Protoctista rather than Protista. All members of this vast phylum have nucleated cells and live in aquatic habitats (freshwater and marine).

Members of the kingdom Protoctista are not animals, which develop from an embryo called a blastula; they are not plants, which develop from an embryo that is not a blastula but is retained in the mother's tissue; they are not fungi which develop from spores and lack cilia and flagella (called undulipodia) at all stages of development; they are not monerans, which have prokaryotic cells.

Fossil protoctists, with thick-walled resting stages or cysts, can be extracted from shale treated with hydrofluoric acid. One of the richest sources of bizarre fossil protoctists was discovered in southern Australia during the late 1950s. Known as the Ediacaran biota, these deposits date back 600 million years ago. Some of these ancient protoctists may have been ancestral to certain animal and plant phyla. In fact, some flattened protoctists discovered in the Ediacaran biota had characteristics resembling lichens. All the Ediacaran biota became extinct by about 530 million years ago and were replaced by shelled Cambrian animals.

Some members of the Kingdom Fungi (in the fungal classes Ascomycetes and Basidiomycetes) are associated with algal cells of the Kingdom Protista (in the algal division Chlorophyta) and/or prokaryotic cyanobacteria of the Kingdom Monera. This complex symbiotic, mutualistic relationship is called lichen. Lichens are essentially lichenised fungi containing unicellular monerans and/or protists.

There are approximately 1.6 million species of living organisms on earth. This number may be much higher because new species are continually being discovered each day, particularly insects and nematodes in remote tropical regions. However, at the present rate of destruction, most of the virgin tropical rain forest will be annihilated by the end of the 20th century, so many species will never be known to humans. It is estimated that 99 per cent of all the species that have ever lived on earth were already extinct before humans ever walked on this planet.

Although humans have a phenomenal impact on the ecology of earth, they are relative newcomers on this great planet. It is estimated that the earth is over 4.5 billion years old, and ancient life forms (such as the cyanobacteria) appeared about 2-3 billion years ago. If the geologic history of the earth is compared to a 24-hour time scale, the first multicellular organisms do not appear until just after 8:00 P.M. and humans are not on the scene until less than a minute before midnight.

There are more than one million species of animals (Kingdom Animalia), more than all the other kingdoms combined. More than half of all animal species are insects (800,000 species), and beetles (300,000 species) comprise the largest order of insects. In fact, if all the species of plants and animals on earth were lined up at random, every 5th species would be a beetle.

Viruses : Viruses do not belong to the above 5 kingdoms of life. They are much smaller and much less complex than cells. They are macromolecular units composed of DNA or RNA surrounded by an outer protein shell. They have no membrane-bound organelles, no ribosomes (organelle site of protein synthesis), no cytoplasm (living contents of a cell), and no source of energy production of their own. They do not exhibit autopoiesis-i.e. they do not have the self-maintenance metabolic reactions of living systems. Viruses lack cellular respiration, ATP-production, gas exchange, etc.

However, they do reproduce, but at the expense of the host cell. Like obligate parasites, they are only capable of reproduction within living cells. In a sense, viruses hijack the host cell and force it to produce more viruses through DNA replication and protein synthesis. Outside of their host cells, viruses can survive as minute macromolecular particles. Viruses may attack animals and plants. Infectious human viruses can be dispersed though the air (airborne viruses) or body fluids (HIV virus). Epidemic viruses (such as HIV) that are passed from person to person via sexual conjugation are remarkably similar to computer viruses.

Unfortunately in humans there is no resident anti-virus programme to alert you of a potential infection, or to quickly scan your body and delete the invader once it has entered your system. Humans must rely on their amazing antibody and cell-mediated immune response, one of the most complex and remarkable achievements in the evolution of living systems. The discovery of a virus called "mimivirus" in 1992 complicates the placement of viruses in the overall classification scheme for living organisms.

Whether mimivirus should be placed in an existing domain (super-kingdom), or in its own domain, remains to be seen. Prior to this discovery, viruses were generally considered non-living until they hijack a living cell. Officially, this virus got its name because it mimics bacteria in size and complexity. Mimivirus was found inside an amoeba within a cooling tower in Bradford, UK. Mimivirus is the largest known virus, about 0.8 micrometers (800 nanometers) across.

In fact it is larger than the bacterium causing gonorrhoea. The virus genome contains 1.2 million bases, more than many bacteria. The bases make up 1,260 genes, which makes it as complex as some bacteria. Most viruses use either DNA or RNA to carry their genetic information, but mimivirus has both of these nucleic acids. In addition, mimivirus can make about 150 of its own proteins, and can even repair its own DNA if it gets damaged. Normal viruses are not capable of protein synthesis or DNA repair on their own, they must rely on the organelles of their host cells for these activities.

The most morphologically and biochemically diverse, non-animal kingdom is the Plantae or Plant Kingdom. It is subdivided into the following 10 phyla or divisions. Note: These names vary considerably, depending on which botany reference you are using.

Kingdom Plantae

Nonvascular Plants : No water-conducting cells (xylem).

- Division Bryophyta (mosses and liverworts).

Vascular plants: Xylem tissue, true roots, stems & leaves.

- Pteridophytes: Spores but no seeds
  - — Division Psilophyta (Psilotum or whisk fern).
  - — Division Lycophyta (club mosses).
  - — Division Sphenophyta (horsetails).
  - — Division Pterophyta (ferns).
  - — Spermatophytes: Seed Plants

Gymnosperms : Naked Seeds

- Division Cycadophyta (cycads).
- Division Ginkgophyta (maidenhair tree).
- Division Gnetophyta (mormon tea & Welwitschia).
- Division Coniferophyta (conifers).

Angiosperms : Seeds Enclosed In A Fruit

- Division Anthophyta (flowering plants)

The complete hierarchal breakdown is Kingdom-Phylum (Division)-Class-Order-Family-Genus-Species. To remember this sequence, the following mnemonic device is often helpful:

King--Phillip--Came--Over--For--Good--Soup

A Biological and Military (Army) Organisational Hierarchy Compared:

Biological Organisation	*Military Organisation*
Kingdom (one or more phyla)	Brigade (two or more regiments)
Phylum (one or more classes)	Regiment (two or more battalions)

Class (one or more orders)	Battalion (two or more companies)
Order (one or more families)	Company (two or more platoons)
Family (one or more genera)	Platoon (two or more squads)
Genus (one or more species)	Squad (a group of 12 soldiers)
Species (a distinct kind or unit)	Soldier (a distinct kind or unit)

The complete taxonomic hierarchy of a marine lichen of the rocky Pacific coast Verrucaria maura with the minute aquatic flowering plant Wolffia borealis:

Kingdom	Fungi	Plantae
Phylum	Eumycota	Tracheophyta
Class	Ascomycetes	Angiospermae
Order	Pyrenulales	Arales
Family	Verrucariaceae	Lemnaceae
Genus	Verrucaria	Wolffia
Species	maura	borealis

The plant kingdom includes non-vascular and vascular plants. Non-vascular plants lack a water-conducting system of tubular cells (called xylem tissue), and do not have true roots, stems and leaves. Like algae and fungi, the plant body of some non-vascular plants is often called a thallus. Non-vascular plants are all placed in the Division Bryophyta, including the mosses and liverworts.

The vast majority of the plant kingdom are vascular, with tubular, water-conducting cells called xylem tissue. Like a microscopic pipeline system, they are arranged end-to-end from the roots to the leaves. Unlike non-vascular plants, they have true roots, stems and leaves. Some references place all the vascular plants in a separate phylum or division called the Tracheophyta.

Most botanists now subdivide vascular plants into 9 divisions. More primitive vascular plants that reproduce by spores, but without seeds, are called pteridophytes, and include the 4 divisions Psilophyta (whisk ferns), Lycophyta (club mosses), Sphenophyta (horsetails), and Pterophyta (ferns). Seed-bearing vascular plants are called spermatophytes and include the primitive gymnosperms (with immature seeds or ovules naked and exposed directly to pollen) and the more advanced angiosperms (with ovules enclosed in an ovary that ripens into a fruit).

Gymnosperms include the 4 divisions Cycadophyta (cycads), Ginkgophyta (maidenhair tree), Gnetophyta (mormon tea & the bizarre South African Welwitschia), and the Coniferophyta (conifers). The angiosperms are placed in the single division Anthophyta which includes all the 'flowering plants and 90 per cent of all the plant kingdom.

Twenty of the more than 100 Pinus species on earth. All of these pines are native to the state of California, USA.

- Monterey Pine (P. radiata),
- Bishop Pine (P. muricata),
- Santa Cruz Island Pine (P. remorata),
- Whitebark Pine (P. albicaulis),
- Limber Pine (P. flexilis),
- Beach Pine (P. contorta),
- Lodgepole Pine (P. murrayana),
- Western White Pine (P. monticola),
- Knobcone Pine (P. attenuata),
- Bristlecone Pine (P. longaeva),
- Foxtail Pine (P. balfouriana),
- Four-Leaf Pinyon (P. quadrifolia),
- Two-Leaf Pinyon (P. edulis),
- One-Leaf Pinyon (P. monophylla),
- Ponderosa Pine (P. ponderosa),
- Coulter Pine (P. coulteri),
- Digger Pine (P. sabiniana),
- Torrey Pine (P. torreyana),

- Jeffrey Pine (P. jeffreyi),
- Sugar Pine (P. lambertiana).

In the Jepson Flora of California, P. remorata is now considered a synonym of P. muricata. Another species called the Washoe Pine (P. washoensis), with cones similar to a miniature Jeffrey Pine, is now recognised for California. In addition, the Beach and Lodgepole Pines are now recognised as subspecies of P. contorta, rather than separate species.

Selection and Genetic Drift

Millions of years ago, cypress woodlands containing one or more ancestral species of the cone-bearing genus Cupressus once dominated vast areas of California. During the past 20 million years, as mountains were uplifted and the climate became increasingly more arid, most of these extensive cypress woodlands vanished from the landscape.

In some areas, the cypress were probably unable to compete with more drought resistant, aggressive species, such as impenetrable chaparral shrubs and desert scrub. Although cypress are fire-adapted with serotinous seed cones that open after a fire, they are vulnerable if the fire interval occurs too frequently, before the trees are old enough to produce a sufficient cone crop. Chaparral shrubs quickly re-sprout after a fast-moving brush fire from well-established subterranean lignotubers. This may explain why some cypress groves occur in very rocky, sterile sites with poor soils where the chaparral shrubs can't compete as well.

Today this fascinating genus is represented by 10 species (or 8 species and 2 subspecies), confined to isolated groves scattered throughout the coastal and inland mountains, from the Mexican border to Oregon. Because some of these populations became isolated into "arboreal islands," gradual genetic changes over millions of years resulted in the present-day species and subspecies. This is somewhat analogous to the evolution of Darwin's finches on the Galapagos Islands.

It is quite likely that natural selection played a role in cypress speciation. Cypress of arid inland mountains and valleys (such as Piute cypress, Macnab cypress, Cuyamaca cypress, and Arizona cypress) have glandular (resinous) foliage and are more drought resistant. Coastal species (such as Monterey cypress, Gowen cypress, Santa Cruz cypress and Mendocino cypress) are generally non-glandular without resin glands on the leaf surfaces. Some phenotypic variability, particularly between different isolated groves of the same species may be due (in part) to genetic drift. These differences include slight variations in foliage, bark characteristics (exfoliating vs. persistent), and the general shape of seed cones.

These differences attributed to genetic drift are analogous to racial differences in people, such as different blood type percentages and facial characteristics. The relatively short period of isolation for Cupressus (cypress) species may be one of the reasons taxonomists disagree on the total number of species native to North America. These numbers have fluctuated greatly in subsequent publications. In addition, the nursery trade has added several cultivated varieties, including at least four different cultivars for the Arizona cypress. New evidence from DNA sequencing has further complicated the number of cypress species, including the transfer of other conifer genera into the genus Cupressus.

For example, the Jepson Manual of California Plants lists ten species; however, two of these C. nootkatensis (Alaska cedar) and C. lawsoniana (Port Orford cedar) were formerly placed in the genus Chamaecyparis. It is possible that some of the isolated species of Cupressus in California and Arizona have not been isolated long enough to warrant the status of a species. In fact, this is why most modern floras have consolidated four species into subspecies of the Arizona cypress (C. arizonica). These species have been isolated long enough for genetic drift to occur, but perhaps not long enough for the development of distinct species populations.

Male (pollen) cones of the Piute cypress (Cupressus

nevadensis) [syn. C. arizonica ssp. nevadensis). Each scalelike leaf bears a dorsal gland that exudes a resin droplet (red arrow). Interior cypress species such as this one typically have glaucous, resinous foliage, presumably an adaptation to dry, arid habitats.

A. Foliage and pollen cones of the Smooth-bark Arizona cypress (Cupressus glabra) [Syn. C. arizonica ssp. glabra]. B. Foliage of the Tecate cypress (C. forbesii). The scalelike leaves of Arizona cypress are glaucous and very glandular (sticky). The scalelike leaves of Tecate cypress are green and without dorsal resin glands.

A grove of Sargent cypress (Cupressus sargentii) in the San Rafael Mountains of Santa Barbara County, California. This species typically grows on outcrops of serpentine in the Coast Ranges of central and northern California. Serpentine is a shiny rock with a waxy luster and feel. It varies in colour from creamy white and shades of green to black. In California, many species of rare and endangered plants are endemic to serpentine outcrops. Genetic drift has undoubtedly occurred in isolated cypress groves such as this one, which are often referred to as "arboreal islands."

Other Members of Division Coniferophyta

Podocarpus gracilior, a member of the Podocarpaceae native to eastern Africa. Although it is sometimes called "fern pine" it does not belong to the genus (Pinus); however, like pines and other cone-bearing species, it does belong to the Division Coniferophyta. The small female cones are composed of 2-4 reduced scales, but usually only one scale bears an ovule that matures into a seed. The seeds have a hard coat surrounded by a fleshy layer (aril).

They are essentially "naked seeds" as in the California nutmeg (Torreya californica) and Pacific yew (Taxus brevifolia), members of the closely-related Yew Family (Taxaceae). Podocarpus seeds are often referred to as fleshy fruits called drupes, but this is incorrect because drupes develop from the ovaries of flowering plants. Another group of conifers with fleshy seed-bearing structures are the junipers (Juniperus) in the Cypress Family (Cupressaceae).

Junipers actually produce small cones with fleshy, fused scales bearing one-several seeds. Podocarpus is a dioecious species, with separate male and female trees in the population. Podocarpus has an ancient lineage dating back to distant relatives that lived during the Jurassic Period 170 million years ago. California nutmeg (Torreya californica), a member of the Division Coniferophyta, Order Taxales, Family Taxaceae. Like Podocarpus, the "naked" seed is enclosed in a fleshy, outer layer (called an aril) which superficially resembles a one-seeded fruit of an angiosperm. The name "nutmeg" is derived from its superficial resemblance to the fruit of the true nutmeg. Pacific yew (Taxus brevifolia), another member of the Division Coniferophyta, Order Taxales, Family Taxaceae that occurs in northern California, Oregon and Washington. Unlike the California nutmeg, the naked seed is not completely enclosed by the fleshy aril. Instead, the seed sits in a cup-shaped aril. Since this species is native to regions of the Pacific northwestern United States containing the timber tree Douglas fir (Pseudotsuga menziesii), it was once considered a weedy species when areas of the forest were logged.

Luckily, the Pacific yew still survives because it is now considered to be an exceedingly valuable species. An extract from the bark (and needles) called taxol has been found to be a very effective treatment for ovarian and breast cancers. It is very important to preserve natural, old growth forests with a diversity of species, some of which may prove to be valuable medicines for the treatment of diseases. The Santa Lucia or bristlecone fir (Abies bracteata) has a tall, slender, steeple-like crown. Seed cones are produced near the top of the slender spire, and they are some of the most unusual cones of all cone-bearing trees on earth. Long, spine-like bracts extend outwardly from between the cone scales, and resemble the antennae of a space satellite.

4

Functions of Animal Body

Physiology (Gr. *Physiologikos* = discourse on natural knowledge) is the study of normal function in living organisms. The term was introduced by the French physician *Jean Femel* in 1552. As the living organisms are composed of chemical substances, physiology involves the application of the laws of chemistry and physics.

Physiology studies not only the functional activities themselves, and their inter-relationship, but it seeks to explore their mechanisms. It defines the characteristics of bodily activities and explores the factors which influence them. It studies the conditions which favour these activities, the mechanisms involved in them and the way in which various processes modify each other.

The field of physiology is often subdivided into 2 divisions : *General Physiology*—the study of fundamental processes in terms of chemistry and physics, and Comparative Physiology—the attempt to discern functional similarities and differences between gross systems of whole organisms.

Functional Structure of the Body

The energy used by the effector muscles and by the mechanisms responsible for maintaining the fabric of the body is derived from food stuffs. A food stuff is a chemical substance which can be taken in or absorbed by a cell and serves it as a source of chemical energy. Many food-stuffs are available to man from both plant and animal sources. They are taken into a long coiled tube, the *gastrointestinal tract,* in appropriate amounts which are normally automatically regulated to the needs of the individual. Once the food-stuffs have entered the digestive tract they are broken down into simpler constituents by the action of enzymes which are formed by glands opening along its length. These are salivary glands, liver pancreas and other intestinals glands. The products of digestion are in a form suitable to pass through the wall of the digestive tract into the blood stream; the unwanted residues are rejected. The nutrients are then carried by the blood to all parts of the body where they are either used for energy, converted into building materials or stored as food reserves for future use.

The liberation of energy from food involves a chemical process *oxidation,* which is analogous to the burning of wood or coal in air with the production of heat, ash and vapour. In the body, oxidation depends upon the activity of enzymes, but it requires oxygen and gives off carbon dioxide and waste products, which must be supplied and removed adequately or the organism rapidly dies. The exchange of gases is the function of the lungs which are spongy organs where gases can pass rapidly to and from air and blood and through which all the blood passes at frequent intervals.

The carbon dioxide is breathed way but the oxygen is carried by the haemoglobin of the blood cells to the tissues, where reverse process of discharge of oxygen and absorption and chemical processes involved are so dovetailed and adjusted that these complex exchanges continue with the utmost precision and efficiency for very long periods. They are self-compensating and adjusting, like all other physiological activities. The elaborate metabolic processes supporting life lead to the production of unwanted *waste products* which have to be excreted if the organism is not to be poisoned by its own products. These are carried by the blood to the kidneys, which are filters with the most accurate powers of removing unwanted material and returning chemicals of value to the body. Small amounts of metabolic wastes are excreted In the breath, the sweat and the faces. The skin also helps in the regulation of water balance in the body.

The ceaseless traffic of gases, food, waste products and chemical messengers, hormones, in the blood stream is made possible by the heart and *circulation*. The heart consists of two pumps made of muscle, each joined to the other through an immensely complex closed system of branching tubes. One sets of tubes runs to and from all the body organs, muscle and brain, and is called the *systemic circulation*. The other set, the *pulmonary circulation* conveys blood to and from the lungs only. The heart pumps steadily away throughout life and the blood flows continuously through these closed circuits. The tissue cells are never touched by the blood itself; all the interchange of gases and chemicals occur by solution and diffusion in the tissue fluid, which permeates the whole body and infiltrates between the cells.

The *lymphoid tissue* consists of lymphoid cells, lymphatic channels and the lymph nodes. It is distributed throughout the body, *e.g.*, lymph nodes, pharyngeal tonsils, spleen, Peyer's patches, vermiform appendix, etc. The cells of the *reticulo-endothelial system* carry out as diverse functions as phagocytosis, formation and destruction of red blood cells, etc. The effector organs, especially the muscles, and the nervous integrate and co-ordinate all other activities of the body. The *muscles* carry out the conscious needs and intentions of the individual but this is only a mall part of their function. They help in movement of the limbs, locomotion and liberation of heat and energy. To a great extent the contraction of muscles is unconscious and automatic, forming what are called reflexes. A *reflex* is an involuntary response of a muscle or gland to impulses conveyed from another part of the body by nervous pathways. Reflex responses may attain elaborate patterns of interlocking relationship such as the mating and breeding behaviour of birds, but they are also responsible for a large series of mundane activities like standing, walking, running, breathing, micturition and defecation.

The *brain* is the site of consciousness, the seat of memory and the organ to which, above all, man owes his supremacy over all other living forms. The faculty of learning is found in animals but only man is known to possess the powers of creative thought and abstract thinking which have led to the development of civilization.

The *autonomic nervous system* consists of *sympathetic* and parasympathetic nerves. Hypothalamus is the higher centre for the autonomic nervous system. Hormones are the chemical secretions of ductless glands like pituitary, thyroid, parathyroids, adrenals, islets of langerhans in the pancrease and the gonads. Hormones circulate in blood stream and help in various metabolic activities. The *sense organs* mainly consist of taste, smell, vision and hearing.

Finally, comes the capacity for *reproduction,* without which all else would be futile. The outlines of the mechanism of fertilization and the elaborate hormonal controls of the profound bodily changes occurring in the mother are fairly clear. The mechanism of heredity and development, the way in which the unparalleled complexity of the adult body derives from the minute speck of nucleoprotein in the fertilized ovum, is still only dimly comprehended and in comparison, the intricate interplay of function in the adult body is a simple problem. The student of physiology is at first rather like a bee outside a hot-house. He sees all sorts of mysterious and exciting flowers inside but every entry is stopped by the glass of unforeseen problems. Each new approach seems blocked by its own difficulty, every description seems to depend upon a new obscurity. A door must be opened somewhere, a start must be made.

5

Rules of Heredity

Gregor Mendel was an Augustinian monk whose experiments with plant breeding conducted at the monastery of St. Thomas in Brunn, Austria between 1854 and 1869 formed the basis of modern genetics. Mendel was not the first to conduct breeding experiments, but he was the first to discover rules governing heredity.

His success was due to his choice of experimental plant (the garden pea, Pisum sativum) and his knowledge of mathematics, especially statistics. The garden pea reproduces by self fertilization and so inbreeding over many generations can detect genetically pure lines. Furthermore, this species possesses several discontinuous traits, i.e., aspects of the organism's outward appearance (phenotype) that exist in only two character states with no intermediates between them.

Each organism possesses one of the two character states. Mendel chose seven discontinuous traits to study: seed shape (either round or wrinkled), seed colour (yellow or green), pod shape (smooth or wrinkled), pod colour (green or yellow), flower colour (red or white), flower position (axial or terminal) and vine length (tall or short). By crossbreeding genetically pure line individuals with contrasting character states for each trait and observing the phenotypes of their offspring, he could determine the pattern of inheritance.

Since these character states are discontinuous (either-or expressions) with no intermediates, he could quantify his results by simply counting the two different phenotypes among the offspring of each experimental cross.

His knowledge of statistics enabled him to recognise that his results followed exact proportions or ratios, even though small sample sise caused variations in observed ratios from the theoretical or expected ratios he was able to recognise.

Mendel started his work by allowing individuals with contrasting character states to self-fertilize through many generations until he was convinced that their traits were pure. Such strains of individuals are termed "true breeding" with respect to the trait in question.

Whenever a true breeding individual is crossed to other true-breeders of its type, the result is that the offspring have the same phenotype as the parents (assuming the environment was functionally the same). He then selected two of these (one with each character state) for crossbreeding.

These two individuals constituted the parental or P generation. Crossbreeding was effected by removing the pollen producing stamens from both plants and then using the pollen of one to fertilize the ovules of the other. The offspring produced by this cross constituted the first filial (or F_1) generation. He then crossbred these individuals or simply let them self-fertilize to produce the second filial (F_2) generation.

Mendel's results can be illustrated by the phenotypic expression of seed shape when he crossed individuals from a pure line of plants with round seeds with individuals from a pure line of plants with wrinkled seeds.

P generation : round x wrinkled	
F_1 generation	all offspring had round seeds
F_2 generation	3:1 ratio of offspring with round and wrinkled seeds (3 round : 1 wrinkled)

When he performed this experiment with the six other traits, he obtained the exact same results. Thus, he recognised a pattern: a 3:1 phenotypic ratio in the F_2 generation. Remember that the purpose of induction in the inductodeductive method of science is to generalise patterns based on a limited number of observations.

Mendel extended his experimental results (based only on the observation of seven traits in a single plant species) through inductive reasoning to claim that all discontinuous traits no matter what animal or plant species is studied would follow the same pattern. The next step was to find a reason why this pattern obtained, i.e., to develop an hypothesis to explain this result.

Hypothesis

To explain this pattern Mendel hypothesised that each discontinuous phenotypic trait in an individual was controlled by two hereditary factors. The alternative hypothesis that each trait was controlled by one factor was not a viable explanation because it could not explain the presence of wrinkled seeds in the F_2 generation. Since all F_1 plants produced only round seeds, then the factor controlling the wrinkled condition would have to have been lost.

But, by postulating two hereditary factors behind each trait, Mendel could explain the reappearance of wrinkled-seeded plants in the second filial generation. To account for the observation that all F_1 individuals had round seeds, Mendel suggested that the factor for wrinkled seeds was masked by the factor for round seeds and so the wrinkled seed factor was recessive to the dominant round seed factor. Whereas dominant factors can influence the phenotype when only one is present, recessive factors must be paired together to produce a phenotypic effect.

It must be emphasised that these hereditary factors (= genes or more precisely, alleles) cannot be observed directly; their existence is only inferred from phenotypes produced in breeding experiments. Mendel explained the hereditary make-up (genotypes = factor pairs) of his experimental plants as follows:

P generation :	*round (RR) x wrinkled (rr)*
F_1 generation	all round (Rr)
F_2 generation	3 round (1 RR + 2 Rr): 1 wrinkled (rr)

To distinguish the three different genotypes in a monohybrid cross (a cross between two individuals who are heterozygous for a single trait) the following terminology is used: R = dominant allele, r = recessive allele, RR = homozygous dominant genotype, Rr = heterozygous genotype, and rr = homozygous recessive genotype.

Law of Segregation

If, as Mendel hypothesised, each discontinuous phenotype is controlled by two alleles, it follows then that only one is passed on to an offspring by each parent, otherwise the number of alleles would double with each generation. This corollary of Mendel's hypothesis is known as his law of segregation that states that the two alleles which together form a genotype segregate into sister gametes when gametes are formed during reproduction.

Note that for a single trait gametes are designated by a single letter (an allele) due to the law of segregation, but genotypes contain two letters because Mendel hypothesised that genotypes or two hereditary factors control a single phenotypic trait.

Use of Deduction

For an hypothesis to be a scientific hypothesis, it must generate testable predictions through the process of deduction. Remember that a prediction is a fact that is deduced from the hypothesis and must be true if the hypothesis is true.

The fact that only phenotypes can be observed provided Mendel with the opportunity to make two predictions about the unseen genotypes in the F_2 generation. The two predictions that follow from Mendel's hypothesis are:

- Underlying the dominant phenotypes in the F_2 generation are two different genotypes: RR and Rr.
- The ratio of these two predicted genotypes is 1 RR to 2Rr. Thus, the 3:1 phenotypic ratio is accompanied by a 1:2:1 genotypic ratio.

According to his law of segregation, the two different factors (R and r) in the F_1 heterozygotes segregated into sister gametes with the result that males produced two types of pollen grains and females produced two types of eggs in their ovules. Since fertilization is a random process, each type of pollen grain has an equal chance of fertilizing each of the two types of eggs (here again is an example of how Mendel's knowledge of mathematics helped him interpret results).

Consequently, the F_2 generation should contain three genotypes in the following ratio: one RR; two Rr; one rr. This expected result is illustrated below with a Punnett Square. A Punnett square is simply a device for listing all possible interactions. It is a nice tool to start your understanding of genetics with, but we will see that it soon becomes cumbersome and we will abandon it. However, for the moment:

	Male Gametes	*(from pollen)*
Female Gametes (from ovule)	R	r
	R RR (round)	Rr (round)
	r Rr (round)	rr (wrinkled)

Hypothesis Testing

Mendel's two predictions could be tested in either of two ways. He could allow all F_2 individuals to self-fertilize, or he could backcross each one with a homozygous recessive plant (test cross). The phenotypic results of the test cross would reveal the presence of the three different genotypes even though the genotypes could not themselves be observed. Each genotype would produce a different set of offspring that would be phenotypically recognisable. The test cross results are presented below:

- — RR x rr = all round (Rr)
- — Rr x rr = both round and wrinkled offspring (50 per cent Rr; 50 per cent rr)
- — rr x rr = all wrinkled (rr)

Other Forms of Heredity

Particulate and Blending Inheritance : Mendel hypothesised a form of particulate inheritance with the hereditary units consisting of free-floating particles. This is in stark contrast to the prevailing idea of his day that inheritance followed a blending pattern, i.e., the parental traits blended together to form the offspring traits. An example of blending inheritance would be the production of pink-flowered offspring from a cross between a red-flowered plant and a whiteflowered one.

Mendel's concept of dominance enabled the particles (alleles) to retain their integrity and be passed on intact to offspring. A particulate inheritance theory was essential for Darwin's theory of natural selection, otherwise any selectively favoured phenotype would be blended away as soon as the selected individual reproduced with one of a different phenotype. Thus, Mendel's theory of inheritance provided the essential requirement for Darwin's theory of evolution: an explanation of heredity based on particulate genes, even though that was not recognised at the time.

On the other hand, there are many examples of inheritance that appear to be blending. This was why the concept existed in the first place. In the example of pink, red and white flowers above, the system would work like this:

P true breeding red$\left(C^R\ C^R\right)$ X true breeding white$\left(C^W\ C^W\right)$ F_1 all pink $\left(C^R\ C^W\right)$

This sort of arrangement is called "incomplete dominance" or "lack of dominance"-neither allele at the "colour" locus is dominant over the other. But notice that unlike earlier non-particulate models of inheritance that the genetic factors do not blend. If these F_1 individuals are crossed with each other we get:

1 red$\left(C^R\ C^R\right)$ 2 pink$\left(C^R\ C^W\right)$ 1 white $\left(C^W\ C^W\right)$

So the factors still exist-as we now know to be true for DNA-inheritance of two factors does not cause them to mix together permanently into some new form.

Qualitative and Quantitative Inheritance : Mendel was successful because he chose discontinuous traits for his analysis and these traits are qualitatively different, i.e., either/or traits with no intermediate conditions. Many traits, however, are quantitative rather than qualitative and vary continuously within a population. Such traits, e.g., height and weight, tend to follow a bellshaped curve when the number of individuals is plotted against the range of the variable trait.

With continuous or quantitative variation there will always be a phenotype in between any two chosen for comparison. The answer is that Mendel's theory explains both qualitative (discontinuous) and quantitative (continuous) patterns of individual variation. With continuous variation, however, many different genes influence the same trait, not just one as Mendel proposed.

Hence, crossing individuals with different character states

of a continuously varying phenotype will not produce the discrete ratios that enabled Mendel to discover his laws of segregation and independent assortment. The genetics of continuous variation is not different from that of discontinuous variation - it is just far more complex.

Chromosomal Basis of Inheritance

A gene is a region of a chromosome (piece of DNA) that contains a set of instructions for synthesising a protein (polypeptide), and that these instructions are passed from cell to cell and from generation to generation as inheritable traits or characteristics. Within an organism, mitosis ensures that each cell formed has the precise set of genetic information of its chromosomes. Meiosis and sexual reproduction are the mechanisms used to pass the genetic instructions from parent to offspring.

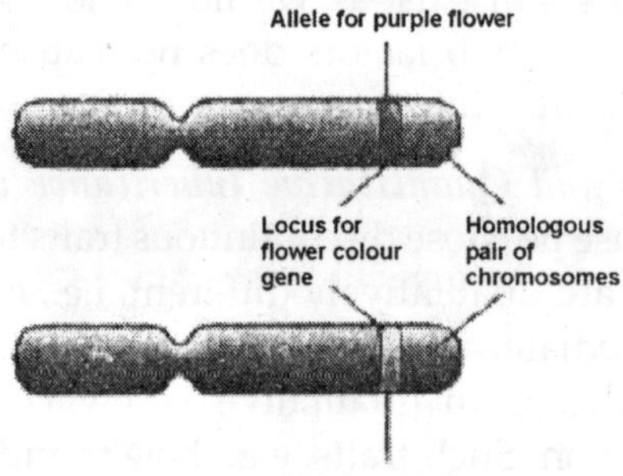

Chromosomes of diploid organisms

The chromosomes of diploid organisms come in pairs, the homologous chromosomes, so that each cell of a diploid organism has two genes for each inheritable trait, one on each of the homologous chromosome pairs. The alternative forms (variants) a gene can have on the homologous chromosomes are called alleles. The precise location where a gene is found on a chromosome is known as the locus.

A gamete has one of each "gene pair" (or one of each homologous chromosome) but not both, and that the diploid number of chromosomes with homologous chromosome pairs, is restored at fertilization, when two gametes fuse. At this time we will look at how genes interact and are expressed in individuals, and how genes are transmitted from generation to generation. We did not always know that genes were located on chromosomes. We did not even know that inheritable traits, or genes, came in pairs.

Mendel in 1865 was the first to state that inheritable traits (genes) came in pairs. Walter Sutton showed that Mendel's principles of inheritance applied to chromosomes and that chromosomes are the units of heredity. The practice of selective animal and plant breeding was well established, virtually nothing was known of the mechanisms of inheritance beyond the presence of an egg and a sperm in animals, and pollen and carpel in plants. It was generally believed at that time that characteristics of parents were "blended" in offspring since offspring generally had features of both parents.

Mendel observed how specific traits of the garden pea were transmitted from generation to generation. Mendel kept precise records of the thousands of offspring (and their characteristics) produced in his crosses. He then established mathematical probabilities and explanations to validate his observations. Although others had studied inheritance, Mendel's educational experiences in math and observing plant variation helped him design and analyse his experiments carefully. Mendel:

- Chose a good organism that had a number of "true breeding" traits easy to observe.
- Designed the experiments carefully. Mendel took plants from true breeding parents (P generation), to first generation (F_1 hybrids), and then self-crossed the first generation offspring to form a second generation (F_2).
- Obtained large sample sizes for good data analysis

Mendel's research led to the following conclusions, two of which are presented as Mendel's Principles:

Inheritance

- There are alternative forms (or variations) of genes, the "units" that determine inherited traits. The alternative forms of a gene are now called alleles. To relate this to what we know about homologous chromosomes, the alleles are located at the same locus on homologous chromosomes. (Specifically, we inherit the alleles for a gene, not the gene).
- An individual will have 2 alleles for each inherited trait. The 2 alleles may be the same, or they may be different. If the two alleles are the same, the individual will be homozygous for that trait. If the two alleles are different, the individual will be heterozygous for the trait.

When the two alleles for a gene pair are different from each other, one will be expressed, and the second will not affect the organism's appearance. The allele always expressed is said to be dominant, and the one that may not be expressed is recessive.

These statements are true for the traits tested in Mendel's peas and for many genes, but are not universally true. Many genes have alleles that are equally expressed, as we shall see, and there are genes that have more than 2 alleles within the population.

- Gametes have just one allele for each trait, because the allele (gene) pairs are separated (or segregated) during meiosis I when homologous chromosomes pair and then separate. 50 per cent of the gametes receive one allele and 50 per cent of the gametes receive the alternative allele when the alleles are heterozygous.

This statement ultimately resulted in Mendel's Principle of Segregation:

Pairs of genes segregate during the formation of gametes (Meiosis), so that each gamete has one of each gene pair (one allele) but not both. Fertilization restores the gene pairs (on the homologous chromosomes). Mendel demonstrated his Principle of Segregation with many monohybrid crosses, looking at one characteristic at a time.

He could further validate his Principle of Segregation with the test cross, a cross in which the F_1 generation, which appeared dominant, was crossed to the recessive parent. Their offspring would exhibit equal proportions of both dominant and recessive forms.

— Mendel's experiments with crossing two traits at one time, a dihybrid cross, resulted in his Principle of Independent Assortment. Each gene pair is distributed (assorts) independently of other gene pairs into gametes during meiosis.

We observed this during meiosis when the homologous pairs of chromosomes align along the equator at metaphase I. Maternal chromosomes of some pairs align towards one pole some of the time, and the other pole some of the time. Each meiosis event has a different alignment pattern.

Gregor Mendel's work was "rediscovered" in 1900 by three independent geneticists who had done studies that came to the same conclusions that Mendel had made. They had the advantage that the processes of mitosis and meiosis were known explaining how genes could be separated.

The next step was accomplished in 1902, when Sutton and Boveri correlated Mendel's conclusions about genes (or inherited traits) to the behaviour of chromosomes during mitosis and meiosis. Sutton is credited with first proposing the chromosome theory of inheritance:

- Chromosomes are in pairs
- Homologous Chromosomes separate during meiosis so that alleles are segregated
- Meiotic products have one of each homologous chromosome but not both
- Fertilization restores the pairs of chromosomes.

Genes Located on Chromosomes

Many human traits follow Mendelian inheritance predictions.

Mendelian inheritance predictions follow the mathematical laws of probability. Although it is fairly "easy" to diagram a monohybrid test and a dihybrid test using Punnett squares, making predictions and looking at results for increasing numbers of genes or other inheritance observations becomes tedious and time consuming. Applying probability laws is much faster and easier.

Some Terms in Mendelian Inheritance Tests

- *True Breeding:*
  — A plant that produces offspring with the same characteristics. The parental generation is a true-breeding generation.
- Cross *Breeding:*
  - A cross between different parental types
  - Offspring produced by cross breeding are called hybrids
- F_1 *Generation*
  — The first generation
  — Generally first generation offspring are bred among themselves to produce the second generation. In Mendel's pea plants they self-fertilized.
- F_2 *Generation*
  — The second generation
  — Mendelian ratios are based on second generation results
- *Punnett Square*
  — A method of visualising Inheritance crosses
- *Gene*
  — The physical unit of heredity; the instructions for

producing a specific characteristic or trait. For example, the characteristic or gene may be flower colour. The alternative forms a gene can have would be the specific flower colours.

— Since diploid organisms have two sets of chromosomes (the homologous chromosome pairs), most "genes" are paired, often called the gene pair

- *Alleles*
  - The alternative forms or variations a gene can have, such as brown or blue for eye colour, or red or white for flower colour.
  - A diploid individual will have two alleles for each gene locus. Within a population there can be more than two alleles for a gene, but only two alleles will be present in any one diploid individual.
- *Locus*
  - The region on a chromosome where a gene is located.
  - The alleles of a gene are located at equivalent places (loci) on the homologous chromosomes
- *Homozygous*
  - The 2 alleles for a gene are the same in an individual
- *Heterozygous*
  - The 2 alleles for a gene are different in an individual
- *Dominant Allele* (loosely and incorrectly called a dominant gene)
  - An allele that is always expressed, whether it is homozygous or heterozygous.
  - A dominant allele masks or covers the expression of its alternative allele.
- *Recessive Allele*
  - An allele which is masked by the presence of its alternative.

— A recessive allele will be expressed only when it is homozygous (when the dominant allele is absent).

- *Phenotype*

— The observable traits of an individual

- *Genotype*

— The specific genetic makeup of an individual, or total combination of alleles present, both those expressed and those not expressed.

You should also review your knowledge of homologous chromosomes and the process of meiosis, since the homologous chromosomes "carry" the alleles, or alternative forms for each gene.

Gene Interactions

The characteristics that Mendel observed had dominant and recessive forms and each inheritable trait was found on different chromosomes.

Early on, some inheritance patterns did not match the expectations proposed by Mendel's principles. Some of the gene interactions that go beyond the basic Mendelian predictions.

Single Gene Variations : Mendel's F_1 offspring always resembled the dominant parent, because each of the genes Mendel chose to study showed complete dominance. When there is no dominant allele, the heterozygote will have a phenotype different from either homozygous form. This is sometimes referred to as an intermediate phenotype.

There are a number of variations in lack of dominance, but each results in heterozygous conditions that have a phenotype different from either homozygous phenotype. In other words, when a gene lacks dominance, there will be three different phenotypes, two homozygous phenotypes (AA and A'A') and a third heterozygous phenotype (AA').

Incomplete dominance

- Failure to completely mask the recessive allele
- The heterozygote first generation has some intermediate phenotype between the two homozygous forms, often a blending of the two alleles.
- Example: snapdragon flowers

Co-Dominance

- Both alleles are equally expressed in the heterozygote-both appear in the heterozygous individuals
- Examples
- Roan Cattle
- Blue Andulusian Fowl
- A-type and B-type Red Blood Cell Coatings

Lack of Dominance is just one of the many different ways that genes are expressed.

Multiple Alleles of One Gene : The "typical" gene has two alleles, one for each of the homologous chromosomes. This is the same for individuals and within the population. For some genes, however, there can be more than two alternative alleles at the single gene locus among members of the species.

- A diploid individual can inherit just two of the possible alleles (one on each of the homologous chromosomes)
- The effects are shown in the study of the population's variation in phenotypes for the gene.
- Example: A,B,O alleles for human rbc coats, the "I" gene.The IA (A) and IB (B) alleles are co-dominant. The i (O) allele is recessive. One consequence of the inheritance of the ABO blood type is that the A and B coatings areantigens, and can trigger antibody reactions in non-complementary individuals. This is important for blood transfusions, but not in genetics.

Genes with More Than One Effect: Pleiotropy (Pleio means

"more") The phenotype conferred by the gene can result in many additional alterations in the individual, many of which seem unrelated to each other. In reality, most genes are probably pleiotropic, but some are dramatically so.

Blood Type	Genotype	Red Blood Cells
AB	AB $I^A I^B$	
A	AA or AO $I^A I^A$ or $I^A i$	
B	BB or BO $I^B I^B$ or $I^B i$	
O	OO ii	

Inheritance of ABO Blood Type

- *Albino condition (No Pigment):* Eye and skin sensitivity to light in many animals
- *Frizzle feathers in chickens:* affects feather shape
- *Feathers can't insulate properly:* Metabolic problems relating to inability to thermoregulate.
- *Abnormal hemoglobin molecule:* affects shape of hemoglobin Protein
- *SRY gene on Y chromosome:* Codes for the protein that activates the genes that code for testes formation. Testes formation activates hormone production that induces development of other male organs.
- *Cystic Fibrosis in Humans:* affects an ion channel protein
- Multiple respiratory problems from mucus blockages
- Blockage of pancreatic ducts
- *Marfan Syndrome:* Fibrillin, a protein needed for connective tissue formation, is defective, leading to "loose" connective tissues and weakness in the skeleton, blood vessels, tendons and ligaments and joint areas.

Interactions Involving More Than One Gene

Controlling Genes : Epistasis (means standing upon or stopping). One gene pair controls or alters the expression of other gene pairs, so that expected phenotypes do not appear. The gene to distribute pigment can be overridden by a second gene that blocks (inhibits) pigment production.

Examples : Mice can have black or brown pigmented fur depending on the inheritance of a gene for pigmentation. Black is dominant and brown, recessive. A second, independent gene prevents the distribution of any pigment in the fur. This gene, when recessive, results in white mice.

In corn, expression of the pigment gene is also controlled by an epistatic gene. Episatic genes can vary in how they control, resulting in different patterns and inheritance ratios. In corn, distribution of pigment requires the epistatic dominant allele. The coat colour and nose and lip pigmentation in Labrador dogs is controlled by two genes, one of which determines distribution of the pigment. Three coat colours are possible: black, brown and yellow. In addition, when both genes are recessive, the nose and lips are also pale, so there are two variants of yellow labs. The pattern is similar to that of the mice pigmentation.

The coat colour and nose and lip pigmentation in Labrador dogs is controlled by two genes, one of which determines distribution of the pigment. Three coat colours are possible: black, brown and yellow. In addition, when both genes are recessive, the nose and lips are also pale, so there are two variants of yellow labs. The pattern is similar to that of the mice pigmentation.

Polygenic Inheritance

Phenotypes resulting from the interaction of one gene or, with epistasis, one gene and a second, controlling gene. Two or more genes can interact to produce greater numbers of phenotypes. When two or more genes interact in ways that result in a number of different phenotypes, we see more variation in the population, with respect to that genetic characteristic. We call this type of inheritance polygenic inheritance. The individual phenotype is the result of the combined interaction of all the alleles at all of the gene loci involved.

Continuous Variation in Polygenic Inheritance : When several copies of a gene interact, continuous variation within the population results. Continuous variation can most easily be demonstrated when population data shows a bell-shaped distribution pattern when graphed. Skin and hair pigmentation and height are two examples of such polygenic inheritance in humans. It is believed that there are at least 3 independent genes, each of which lacks dominance, responsible for producing the melanin pigment in human skin (and in hair).

Chromosomes and Inheritance : Today we know that each cell contains several thousand genes, and that genes are specific regions of chromosomes. Further we know that entire chromosomes, not individual genes, are transmitted by meiosis to the gametes. But the inheritance patterns discussed to date have involved genes located on different chromosomes, so they have followed Mendel's Principle of Independent Assortment. Each homologous chromosome pair assorts independently, not specific genes. Two genes that are located on the same chromosome will be inherited together, and not assort independently during meiosis.

The inheritance of genes that are located on the same chromosome, the subject of gene linkage. This adds some interesting complications to the predicted patterns of inheritance. In 1908, researchers discovered a dihybrid cross in sweet peas that did not give the predicted Mendelian ratio of 9:3:3:1. They could not explain why their results were closer to 75 per cent and 25 per cent (the 3:1 ratio expected for a monohybrid cross). Ultimately it was shown that the flower colour and pollen length (the genes observed) were on the same chromosome. Since we inherit entire chromosomes rather than independent genes, all genes on one chromosome are inherited together as a single unit (called the linkage group), and we should expect a 3:1 inheritance ratio for the linkage group. This was just the first time someone had seen this.

Effect of Recombination on Gene Linkage and Inheritance: Crossing over during meiosis results in the exchange of bits and pieces of DNA between homologous pairs of chromosomes at the chiasmata during prophase I of meiosis. This process of recombination results in gametes (or meiotic products) that are not identical; some of the linkage groups have been changed by the crossing-over. As a result of recombination, new allele combinations are formed, and we have more genetic variation.

Sex-determining Chromosomes and Linkage : One of the earliest discoveries about gene linkage related to another significant thing about chromosomes and species, especially animal species. By the early 1900's it was known that males and females of most species have one pair of "not-exactly-matching" homologous chromosomes, which determined the gender of the individual. These chromosomes were called the sex chromosomes. With the gender-determining chromosomes, one sex, usually female, will have two matching chromosomes (XX) and the other sex will have two unmatched chromosomes (XY). At meiosis, all eggs will contain an X chromosome, but half the sperm gametes will have a Y chromosome and the other half will have an X chromosome.

Some species have the reverse pattern of sex chromosomes (male = XX and female = XY), and some species have one gender (female) with a pair of chromosomes and one gender (male) with a single unmatched chromosome. In all cases the gender with the dissimilar pattern will determine the gender of the offspring.

6

History of the Planet

The creation of the Planet Earth has imposed a fascinating quest in the human beings from the time immemorial. The deeper reality of the creation of the Earth is yet to be opened. Wrapped in the mysterious pages, the subject imposes different suggestive items which are the interest of great scientific and Philosophical studies.

In the very beginning of earth's history, this planet was a giant, red hot, roiling, boiling sea of molten rock - a magma ocean. The heat had been generated by the repeated high speed collisions of much smaller bodies of space rocks that continually clumped together as they collided to form this planet. As the collisions tapered off the earth began to cool, forming a thin crust on its surface.

As the cooling continued, water vapour began to escape and condense in the earth's early atmosphere. Clouds formed and storms raged, raining more and more water down on the primitive earth, cooling the surface further until it was flooded with water, forming the seas. It is theorised that the true age of the earth is about 4.6 billion years old, formed at about the same time as the rest of our solar system.

The oldest rocks geologists have been able to find are 3.9 billion years old. Using radio metric dating methods to determine the age of rocks means scientists have to rely on when the rock was initially formed (as in - when its internal minerals first cooled). In the infancy of our home planet the entire earth was a molten (melted) rock, a magma ocean. Since we can only measure as far back in time as we had solid rock on this planet, we are limited in how we can measure the real age of the earth.

Due to the forces of plate tectonics, our planet is also a very dynamic one; new mountains forming, old ones wearing down, volcanoes melting and reshaping new crust. The continual changing and reshaping of the earth's surface that involves the melting down and reconstructing of old rock has pretty much eliminated most of the original rocks that came with earth when it was newly formed. So the age is a theoretical age.

Mystery of Life

Scientists are still trying to unravel one of the greatest mysteries of earth: When did "life" first appear and how did it happen? It is estimated that the first life forms on earth were primitive, one-celled creatures that appeared about 3 billion years ago. That's pretty much all there was for about the next two billion years. Then suddenly those single celled organisms began to evolve into multicellular organisms. Then an unprecedented profusion of life in incredibly complex forms began to fill the oceans.

Some crawled from the seas and took residence on land, perhaps to escape predators in the ocean. A cascading chain of new and increasingly differentiated forms of life appeared all over the planet, only to be virtually annihilated by an unexplained mass extinction. It would be the first of several mass extinctions in Earth's history. Scientists have been looking increasingly to space to explain these mass extinctions that have been happening almost like clockwork since the beginning of "living" time.

Perhaps we've been getting periodically belted by more space rocks (i.e. asteroids), or the collision of neutron stars happening too close for comfort? Each time a mass extinction occurred, life found a way to come back from the brink. Life has tenaciously clung to this small blue planet for the last three billion years. Scientists are finding new cues as to how life first began on earth in some really interesting places - the deep ocean.

Scientists have studied rocks using radio metric dating methods to determine the age of earth. Another really cool thing they've found in rocks that tells us more about the story of earth's past are the remains of living creatures that have been embedded in the rocks for all time. We call these fossils. It has been the careful study of earth's fossil record that has revealed the exciting picture about the kinds of creatures that once roamed this planet.

Fossilised skeletons of enormous creatures with huge claws and teeth, ancient ancestors of modern day species (such as sharks) that have remained virtually unchanged for millions of years, and pre-historic jungles lush with plant life, all point to a profusion of life and a variety of species that continues to populate the earth, even in the face of periodic mass extinctions.

By studying the fossil record scientists have determined that the earth has experienced very different climates in the past. In fact, general climactic conditions, as well as existing species, are used to define distinct geologic time periods in earth's history. For example, periodic warming of the earth - during the Jurassic and Cretaceous periods - created a profusion of plant and animal life that left behind generous organic materials from their decay. These layers of organic material built up over millions of years undisturbed.

They were eventually covered by younger, overlying sediment and compressed, giving us fossil fuels such as coal, petroleum and natural gas. Alternately, the earth's climate has also experienced periods of extremely cold weather for such prolonged periods that much of the surface was covered in thick sheets of ice. These periods of geologic time are called ice ages and the earth has had several in its history.

Entire species of warmer-climate species died out during these time periods, giving rise to entirely new species of living things which could tolerate and survive in the extremely cold climate. Believe it or not, humans were around during the last ice age - the Holocene (about 11,500 years ago) - and we managed to survive. Creatures like the Woolly Mammoth - a distant relative of modern-day elephants - did not.

Research into the origin of life is the modern incarnation of the ancient concept of abiogenesis. Abiogenesis, in its most general sense, is the generation of life from non-living matter. The term is primarily used in the context of biology and the origin of life. Abiogenesis was long considered to be a very common occurrence until the Law of Biogenesis (omne vivum ex ovo or "all life from other life") became firmly established in modern biology.

The modern definition of abiogenesis is concerned with the formation of the simplest forms of life from primordial chemicals. This is significantly different from the concept of Aristotelian abiogenesis, which postulated the formation of complex organisms. There is no truly "standard" model of the origin of life, however most currently accepted models build in one way or another upon the following discoveries, which are listed in a rough order of postulated emergence:

- Plausible pre-biotic conditions result in the creation of certain basic small molecules (monomers) of life, such as amino acids.
- Phospholipids (of an appropriate length) can spontaneously form lipid bilayers, one of the two basic components of a cell membrane.
- The polymerisation of nucleotides into random RNA molecules might have resulted in self-replicating ribozymes (RNA world hypothesis).

The origin of the basic biomolecules, while not settled, is less controversial than the significance and order of steps 2 and 3. The basic inorganic chemicals from which life was formed are methane (CH_4), ammonia (NH_3), water (H_2O), hydrogen sulphide (H_2S), carbon dioxide (CO_2), and phosphate (PO_4). As of 2004, no one has yet synthesised a "protocell" using basic components which has the necessary properties of life (the so-called "bottom-up-approach").

Without such a proof-of-principle, explanations have tended to be short on specifics. However, some researchers are working in this field, notably Jack Szostak at Harvard. Others have argued that a "top-down approach" is more feasible. One such approach attempted by Craig Venter and others at The Institute for Genomic Research involved engineering existing prokaryotic cells with progressively fewer genes, attempting to discern at which point the most minimal requirements for life were reached.

When the earth formed some 4.6 billion years ago, it was a lifeless, inhospitable place. A billion years later it was teeming with organisms resembling blue-green algae. This long-standing question continues to generate fascinating conjectures and ingenious experiments, many of which centre on the possibility that the advent of self-replicating RNA was a critical milestone on the road to life.

Before the mid-17th century, most people believed that God had created humankind and other higher organisms and that insects, frogs and other small creatures could arise spontaneously in mud or decaying matter. For the next two centuries, those ideas were subjected to increasingly severe criticism, and in the mid-19th century two important scientific advances set the stage for modern discussions of the origin of life.

The second advance, the theory of natural selection, suggested an answer. According to this proposal, set forth by Charles Darwin and Alfred Russel Wallace, some of the differences between individuals in a population are heritable.

When the environment changes, individuals bearing traits that provide the best adaptation to the new environment meet with the greatest reproductive success.

Consequently, the next generation contains an increased percentage of well-adapted individuals displaying the helpful characteristics. In other words, environmental pressures select adaptive traits for perpetuation. Repeated generation after generation, natural selection could thus lead to the evolution of complex organisms from simple ones.

The theory therefore implied that all current life-forms could have evolved from a single, simple progenitor - an organism now referred to as life's last common ancestor. This life-form is said to be "last" not "first" because it is the nearest shared ancestor of all contemporary organisms; more distant ancestors must have appeared earlier. Darwin, bending somewhat to the religious biases of his time, posited in the final paragraph of The Origin of Species that "the Creator" originally breathed life "into a few forms or into one."

Then evolution rook over: "From so simple a beginning endless forms most beautiful and most wonderful have been, and are being evolved." In private correspondence, however, he suggested life could have arisen through chemistry, "in some warm little pond, with all sorts of ammonia and phosphoric salts, light, heat, electricity, etc. present."

For much of the 20th century, origin-of-life research has aimed to flesh out Darwin's private hypothesis - to elucidate how, without supernatural intervention, spontaneous interaction of the relatively simple molecules dissolved in the lakes or oceans of the pre-biotic world could have yielded life's last common ancestor. Finding a solution to this problem requires knowing something about that ancestor's characteristics.

Obviously, it had to possess genetic information - that is, heritable instructions for functioning and reproducing -and the means to replicate and carry out those instructions.

Otherwise it would have left no descendants: Also, its system for replicating its genetic material had to allow for some random variation in the heritable characteristics of the offspring so that new traits could be selected and lead to the creation of diverse species.

Scientists have attained more insight into the character of the last common ancestor by identifying commonalities in contemporary organisms. One can safely infer that intricate features present in all modern varieties of life also appeared in that common ancestor. After all, it is next to impossible for such universal traits to have evolved separately. The rationale is the same as would apply to the discovery of two virtually identical screenplays, differing only by a few words.

It would be unreasonable to think that the scripts were created independently by two separate authors. By the same token, it would be safe to assume that one script was an imperfect replica of the other or that both versions were slightly altered copies of a third. One readily apparent commonality is that all living things consist of similar organic (carbon-rich) compounds. Another shared property is that the proteins found in present-day organisms are fashioned from one set of 20 standard amino acids.

These proteins include enzymes (biological catalysts) that are essential to development, survival and reproduction. Further, contemporary organisms carry their genetic information in nucleic acids - RNA and DNA - and use essentially the same genetic code. This code specifies the amino acid sequences of all the proteins each organism needs. More precisely, the instructions take the form of specific sequences of nucleotides, the building blocks of nucleic acids. These nucleotides consist of a sugar (deoxyribose in DNA, and ribose in RNA), a phosphate group and one of four different nitrogen-containing bases.

In DNA, the bases are adenine (A), guanine (G), cytosine (C) and thymine (T). In RNA, uracil (U) substitutes for thymine.

The bases constitute the alphabet, and triplets of bases form the words. As an example, the triplet CUU in RNA instructs a cell to add the amino acid leucine to a growing strand of protein. From such findings we can infer that our last common ancestor stored genetic information in nucleic acids that specified the composition of all needed proteins.

It also relied on proteins to direct many of the reactions required for self-perpetuation. Anyone trying to solve this puzzle immediately encounters a paradox. Nowadays nucleic acids are synthesised only with the help of proteins, and proteins are synthesised only if their corresponding nucleotide sequence is present. It is extremely improbable that proteins and nucleic acids, both of which are structurally complex, arose spontaneously in the same place at the same time.

Yet it also seems impossible to have one without the other. And so, at first glance, one might have to conclude that life could never, in fact, have originated by chemical means. RNA might well have come first and established what is now called the RNA world - a world in which RNA catalysed all the reactions necessary for a precursor of life's last common ancestor to survive and replicate.

Also posited that RNA could subsequently have developed the ability to link amino acids together into proteins. If pre-biotic RNA had two properties not evident today: a capacity to replicate without the help of proteins and an ability to catalyse every step of protein synthesis. There were a few reasons why we favoured RNA over DNA as the originator of the genetic system, even though DNA is now the main repository of hereditary information. One consideration was that the ribonucleotides in RNA are more readily synthesised than are the deoxyribonucleotides in DNA.

Moreover, it was easy to envision ways that DNA could evolve from RNA and then, being more stable, take over RNA's role as the guardian of heredity. RNA came before proteins in part because we had difficulty composing any scenario in which proteins could replicate in the absence of nucleic acids. During the past 10 years, a fair amount of evidence has lent credence to the idea that the hypothetical RNA world did exist and lead to the advent of life based on DNA, RNA and protein.

Until then, proteins were thought to carry out all catalytic reactions in contemporary organisms. Indeed, the term "enzyme" is usually reserved for proteins. The first ribozymes identified could do little more than cut and join pre-existing RNA. Nevertheless, the fact that they behaved like enzymes added weight to the notion that ancient RNA might also have been catalytic.

Origin of Organic Molecules

The Miller-Urey experiment attempts to recreate the chemical conditions of the primitive Earth in the laboratory, and synthesised some of the building blocks of life. The "Miller experiments" are performed under simulated conditions resembling those thought at the time to have existed shortly after Earth first accredit from the primordial solar nebula. The experiment used a highly reduced mixture of gases (methane, ammonia and hydrogen).

However, it should be noted that the exact composition of the pre-biotic atmosphere of earth is currently somewhat controversial. Other less reducing gases produce a lower yield and variety. It was once thought that appreciable amounts of molecular oxygen were present in the pre-biotic atmosphere, which would have essentially prevented the formation of organic molecules; however, the current scientific consensus is that such was not the case. The experiment showed that some of the basic organic monomers (such as amino acids) that form the building blocks of modern life can be formed spontaneously. Simple organic molecules are of course a long way from a fully functional self-replicating life form; however, in an environment with no pre-existing life these molecules may have accumulated and provided a rich environment for chemical evolution ("soup theory").

On the other hand, the spontaneous formation of complex polymers from abiotically generated monomers under these conditions is not at all a straightforward process. Further, according to Brooks and Shaw (1973), there is no evidence in the geological record that any soup existed. "If there ever was a primitive soup, then we would expect to find at least somewhere on this planet either massive sediments containing enormous amounts of the various nitrogenous organic compounds, acids, purines, pyrimidines, and the like; or in much metamorphosed sediments we should find vast amounts of nitrogenous cokes.

In fact no such materials have been found anywhere on earth." Other sources of complex molecules have been postulated, including sources of extra-terrestrial stellar or interstellar origin. For example, from spectral analyses, organic molecules are known to be present in comets and meteorites. In 2004, a team detected traces of polycyclic aromatic hydrocarbons (PAH's) in a nebula, the most complex molecule, to that date, found in space.

It can be argued that the most crucial challenge unanswered by this theory is how the relatively simple organic building blocks polymerise and form more complex structures, interacting in consistent ways to form a protocell. For example, in an aqueous environment hydrolysis of oligomers/polymers into their constituent monomers would be favoured over the condensation of individual monomers into polymers. Also, the Miller experiment produces many substances that would undergo cross-reactions with the amino acids or terminate the peptide chain. A possible answer to this polymerisation conundrum was provided in 1980s by Gunter Wachtershauser, in his iron-sulphur world theory. In this theory, he postulated the evolution of (bio)chemical pathways as fundamentals of the evolution of life. Moreover, he presented a consistent system of tracing today's biochemistry back to ancestral reactions that provide alternative pathways to the synthesis of organic building blocks from simple gaseous compounds.

In contrast to the classical Miller experiments, which depend on external sources of energy, "Wachtershauser systems" come with a built-in source of energy, sulphides of iron and other minerals (e.g. pyrite). The energy released from redox reactions of these metal sulphides is not only available for the synthesis of organic molecules, but also for the formation of oligomers and polymers. It is therefore hypothesised that such systems may be able to evolve into auto catalytic sets of self-replicating, metabolically active entities that would predate the life forms known today.

The experiment as performed, produced a relatively small yield of dipeptides (0.4-12.4 per cent) and a smaller yield of tripeptides (0.003 per cent) and the authors note that: "under these same conditions dipeptides hydrolysed rapidly." Another criticism of the result is that the experiment did not include any organomolecules that would most likely cross-react or chain-terminate.

The latest modification of the iron-sulphur-hypothesis has been provided by William Martin and Michael Russell in 2002. According to their scenario, the first cellular life forms may have evolved inside so-called black smokers at seafloor spreading zones in the deep sea. These structures consist of microscale caverns that are coated by thin membraneous metal sulphide walls. Therefore, these structures would solve several critical points of the "pure" Wachtershauser systems at once:

- The micro-caverns provide a means of concentrating newly synthesised molecules, thereby increasing the chance of forming oligomers;
- The steep temperature gradients inside a black smoker allow for establishing "optimum zones" of partial reactions in different regions of the black smoker (e.g. monomer synthesis in the hotter, oligomerisation in the colder parts);
- The flow of hydrothermal water through the structure provides a constant source of building blocks and energy (freshly precipitated metal sulphides);

- The model allows for a succession of different steps of cellular evolution (pre-biotic chemistry, monomer and oligomer synthesis, peptide and protein synthesis, RNA world, ribonucleoprotein assembly and DNA world) in a single structure, facilitating exchange between all developmental stages;
- Synthesis of lipids as a means of "closing" the cells against the environment is not necessary, until basically all cellular functions are developed.

This model locates the "last universal common ancestor" (LUCA) inside a black smoker, rather than assuming the existence of a free-living form of LUCA. The last evolutionary step would be the synthesis of a lipid membrane that finally allows the organisms to leave the microcavern system of the black smokers and start their independent lives.

This postulated late acquisition of lipids is consistent with the presence of completely different types of membrane lipids in archaebacteria and eubacteria (plus eukaryotes) with highly similar cellular physiology of all life forms in most other aspects.

Another unsolved issue in chemical evolution is the origin of homochirality, i.e. all monomers having the same "handedness" (amino acids being left handed, and nucleic acid sugars being right handed). Homochirality is essential for the formation of functional ribozymes (and probably proteins too). The origin of homochirality might simply be explained by an initial asymmetry by chance followed by common descent.

Organic Molecules to Protocells

However, there are many different hypotheses regarding the path that might have been taken. Some of these postulate the early appearance of nucleic acids ("genes-first") whereas others postulate the evolution of biochemical reactions and pathways first ("metabolism-first"). Recently, trends are emerging to create hybrid models that combine aspects of both.

RNA World Hypothesis : Genes first models are RNA. The RNA world hypothesis, for example, suggests that relatively short RNA molecules could have spontaneously formed that were capable of catalysing their own continuing replication. Early cell membranes could have formed spontaneously from proteinoids, protein-like molecules that are produced when amino acid solutions are heated.

Other possibilities include systems of chemical reactions taking place within clay substrates or on the surface of pyrite rocks. At this time however, these various hypotheses have incomplete evidence supporting them. Many of them can be simulated and tested in the lab, but a lack of undisturbed sedimentary rock from that early in Earth's history leaves few opportunities to determine what may have actually happened in reality. At this time however, no pre-biotically plausible experiment has confirmed this assumption.

Further, recent experiments suggest that the original estimates of the size of an RNA molecule capable of self-replication were most probably vast underestimates. RNA itself does not appear to be a pre-biotically plausible molecule; therefore, more-modern forms of the RNA World theory propose that a simpler molecule was capable of self-replication.

Iron-sulphur World : Several models reject the idea of the self-replication of a "naked-gene" and postulate the emergence of a primitive metabolism which could provide an environment for the later emergence of RNA replication. One of the earliest incarnations of this idea was put forward in 1924 with Alexander Oparin's notion of primitive self-replicating vesicles which predated the discovery of the structure of DNA.

More abstract and theoretical arguments for the plausibility of the emergence of metabolism without the presence of genes include a mathematical model introduced by Freeman Dyson in the early 1980s. However, the idea that a closed metabolic cycle, such as the reductive citric acid cycle proposed by Gunter Wachtershauser, could form spontaneously remains unsupported. Further, according to Leslie Orgel, a leader in origin-of-life studies for the past several decades, there is reason to believe the assertion will remain so. There is at present no reason to expect that multi-step cycles such as the reductive citric acid cycle will self-organise on the surface of FeS/ FeS2 or some other mineral.

Hybrid Models : A growing realisation of the inadequacy of either pure "genes-first" or "metabolism-first" models is leading the trend towards models that incorporate aspects of each.

Clay Theory

Clay theory postulates complex organic molecules arising gradually on a pre-existing, non-organic replication platform-silicate crystals in solution. Complexity in companion molecules developed as a function of selection pressures on types of clay crystal is then exacted to serve the replication of organic molecules independently of their silicate "launch stage". Cairns-Smith is a staunch critic of other models of chemical evolution. However, he admits, that like many models of the origin of life, his own also has its shortcomings.

Deep-hot Biosphere Model of Gold : A controversial theory put forward by Thomas Gold in the 1990s has life first developing not on the surface of the earth, but several kilometres below the surface. It is now known that microbial life is plentiful up to five kilometres below the earth's surface in the form of archaea, which are generally considered to have originated around the same time or earlier than bacteria, most of which live on the surface including the oceans.

It is claimed that discovery of microbial life below the surface of another body in our solar system would lend significant credence to this theory. He also noted that a trickle of food from a deep, unreachable, source promotes survival because life arising in a puddle of organic material is likely to consume all of its food and become extinct.

Primitive Extraterrestrial Life : An alternative to Earthly abiogenesis is the hypothesis that primitive life may have originally formed extraterrestrially. Organic compounds are relatively common in space, especially in the outer solar system where volatiles are not evaporated by solar heating. Comets are encrusted by outer layers of dark material, thought to be a tar-like substance composed of complex organic material formed from simple carbon compounds after reactions initiated mostly by irradiation by ultraviolet light.

It is supposed that a rain of cometary material on the early Earth could have brought significant quantities of complex organic molecules, and that it is possible that primitive life itself may have formed in space was brought to the surface along with it. A related hypothesis holds that life may have formed first on early Mars, and been transported to Earth when crustal material was blasted off of Mars by asteroid and comet impacts to later fall to Earth's surface.

Theories on the Origin of Life

The leading theories on the origin of life could be divided into three main groups:

- The RNA world
- Under high temperature and pressure
- Extraterrestrial origin

Some background information.

The subject matter is generally divided into five stages:

- Synthesis of organic compounds
- Synthesis of biochemical substances.
- Having to do with the production of large molecules such as proteins.
- Having to do with organised cellular structures.
- Concerns the evolution of macromolecules and metabolism.

The macromolecules of life.

- Amino acids are the building blocks of proteins, which are synthesised at one of many small cellular bodies called ribosomes. In meteorites we can detect over 70 amino acids, but human bodies only make use of 20 amino acids.
- Proteins are organic compounds, which are essential biomolecules of all living organisms. Their elements are: hydrogen, carbon, oxygen, nitrogen and sulphur. They are made up of a series of amino acids
- The nitrogenous base compounds U, C, G, A and T arering compounds which are constituents of nucleic acids.
- Nucleic acids are organic molecular structures consisting of five-carbon sugars, a phosphate and mainly one of the five bases
- Lipids are a wide group of organic compounds having in common their solubility in organic solvents, such as alcohols. They are the constituents of cell membrane and have a multitude of other important roles.

To create macromolecules of life:

Precursor molecule	Macromolecule of life
Formaldehyde CH_2O	Ribose, glycerol
Carbon monoxide + hydrogen $CO + H_2$	Fatty acids
Hydrogen cyanide HCN	Purines (adenine, guanine)
Cyanamide H_2NCN	Peptides and Phospholipids

- The Earth was thought to coalesce about 4.5 billion years ago.
- Microfossils found in ancient rocks from Australia and South Africa demonstrate that terrestrial life was flourishing by 3.5 billion years ago. Even older rocks from Greenland, 3.9 billion years old, contain isotopic finger-prints of carbon that could have belonged to a living organism. In other words, only 100 million years

or so after the earliest possible point when Earth could have supported life, organisms were already well enough established that evidence of them remains today.

- 'Zircon' are minerals of great resistance to high temperature and pressure. Specific analysis of the isotopes of elements (uranium and lead) present in a grain of zircon from the Narryer Gneiss Complex in Western Australia are consistent with the presence of continental crust and even liquid water between 4.4 and 4.3 billions years ago.

It is known that heavy oxygen is produced by the interaction between rock and liquid water, a process that should occur at sufficiently low temperatures. The presence of the heavy isotope of oxygen in the zircons from the sample mentioned above suggest that magma had been produced by an episode of high temperature (or heavy pressure) on the surface of the early Earth, on which there was liquid water.

Prebiotic Earth :

- According to the accretion model, several lunar -(0.01 Earth mass) to Mars-sized (0.1 Earth mass) objects would have impacted the Earth. The number, size and position of the terrestrial planets are partly a consequence of the randomness associated with the small number of these bodies.

The conditions associated with a very large impactor of lunar or Martian size are extreme and would have wiped out much of the earlier structure of the planet to depth of many kilometres. Any life forms that may have existed at the time would not have survived. It is estimated to be 4.5 billion years ago, the age of the oldest moon rock.

- The size distribution and the flux of the objects that formed the Earth in the first 10 years are important. With a continual flux of smaller objects, water would have been kept in the atmosphere and hydrogen would have been lost to space at a significant rate. The water that dissolved in the molten rock would have been transported to the interior by convection.

Below about 100km depth, the water would have been lost by reaction with metallic iron to form iron hydride in solution in the iron phase and ferrous oxide in silicates. Carbon as CO, CO_2, or CH_4 is not readily lost to space, but carbon is soluble in iron and some should have entered the core.

- Accretion and the formation of the Earth's core are thought to have occurred simultaneously. Giant impacts supplied their already differentiated cores to the Earth. Enormous amounts of heat were added to the Earth by the processes. Most of this heat had to escape to space for the Earth to cool below the melting point. Plate tectonics is much too slow to release this heat. A more likely form of convection is by total melting of material at depth and eruption of the material to the surface. As a result, pre-existing metallic iron carried by accreting bodies would have been removed from near-surfaced regions.
- If segregation of iron into the core was inefficient and metal remained in the upper mantle to buffer the redox state of magmas, or if the metallic core, mantle, and outgassed volatiles were all in thermodynamic equilibrium during the period of rapid heat loss, CO and CH_4 would have been the predominant thermodynamically stable forms of carbon injected into the atmosphere form the interior. The primitive atmosphere would have been rich in H_2, CO, CH_4, NH_3 and H_2S.

Furthermore, the impacts of large iron-bearing asteroids comparable in mass to that of the oceans may have yielded pulses of highly reduced gases as a result of equilibration between vaporised iron and elemental hydrogen, carbon, nitrogen and sulphur.

- In the absence of metal, the composition of the

> atmosphere would have been determined by the redox state of the near-surface metal-free silicate melts. Under this circumstance (neutral redox composition), CO_2 would have been the predominant carbon-bearing gas; other major atmospheric constituents would have been N_2 and H_2O; sulphur would have emerged from hydrothermal vents as H_2S; and CH_4. CO and H_2 would have occurred at trace amounts at best.

The non-enzymatic synthesis of biological monomers in an atmosphere of methane, nitrogen, ammonia, and water: There are many observations that gaseous mixtures, for example, methane, nitrogen, ammonia, and water, if supplied if energy such as spark discharges, produce amino acids including those found regularly in proteins.

The distribution if monomers so produced is qualitatively and quantitatively similar to that found in carbonaceous meteorites. In addition, most protein amino acids may be produced non-enzymatically starting with simple organic compounds such as formaldehyde and hydroxylamine.

Furthermore, the abiotic routes of formation of all the components of DNA and RNA are known. Sugars easily form spontaneously from formaldehyde; polymerisation occurs under alkaline conditions. The condensation of hydrogen cyanide in the presence of ammonia produces amino acids as well as the purine nucleotide bases, adenine and guanine, components of all nucleic acids. Cytosine, a base found in nucleic acids, can be readily synthesised from cyanoacetylene.

By deamination, cytosine yields another major base of RNA, uracil. Thymine, a major base of DNA, which in today's genetic code is informationally equivalent to uracil, can be formed from the condensation of uracil with formaldehyde. In the presence of phosphate the phosphorylated forms of the nucleotides of these bases can be produced non-enzymatically. Fatty acids may be formed from carbon monoxide and hydrogen in the presence of nickel-iron catalysts, catalysts that might have been brought in by meteorites. Glycerol is a component of fats that has also been obtained non-enzymatically in the laboratory by reduction of glyceraldehyde. Glyceraldehyde itself, a common intermediate in cell energy-yielding reactions, may be formed by condensation of formaldehyde under alkaline conditions.

The RNA world :

- Darwin's theory on natural selection implied that all current life-forms could have evolved from a single, simple progenitor which is referred to as the 'last' common ancestor. By identifying the commonalities in contemporary organisms, we can infer that intricate features present in all modern varieties of life also appeared in the common ancestor because it is next to impossible for such universal traits to have evolved separately.

  One feature is the presence of genetic information and the means to replicate and carry out those heritable instructions for functioning and reproducing. In addition, the system for replicating genetic material had to allow for some random variation in the heritable characteristics of the offspring so that new traits could be selected and lead to the creation of diverse species.
- Another commonality is that all living things consist of similar organic (carbon-rich) compounds. Also, the proteins found in present-day organisms are fashioned from one set of 20 standard amino acids. Furthermore, all contemporary organisms carry their genetic information in nucleic acids - RNA and DNA - and use essentially the same genetic code.

From these findings, we can infer that the last common ancestor stored genetic information in nucleic acids that specified the composition of all needed proteins. It also relied on proteins to direct many of the reactions required for self perpetuation. The question now becomes where does the proteins and nucleic acids come from?

The Chicken and Egg problem (nucleic acids and proteins) Nowadays nucleic acids are synthesised only with the help of proteins, and proteins are synthesised only if their corresponding nucleotide sequence is present. It is extremely improbable that proteins and nucleic acids, both of which are structurally complex, arose spontaneously in the same place at the same time. Yet it also seems impossible to have one without the other.

The paradox is resolved by the RNA world - a world in which RNA catalysed all the reactions necessary for the precursor of life's last common ancestor to survive and replicate. The RNA could subsequently have developed the ability to link the amino acids together into proteins. For the RNA world to exist, RNA needs the capacity to replicate without the help of proteins and an ability to catalyse every step of protein synthesis.

- Why RNA is favoured over DNA as the originator of the genetic system: The ribonucleotides in RNA are more readily synthesised than are the deoxyribonucleotides in DNA. It is easy to envision ways that DNA could evolve from RNA and then, being more stable, take over RNA's role as the guardian of heredity. Researchers suspect that RNA came before proteins because they face difficulty composing any scenario in which proteins could replicate in the absence of nucleic acids.
- In 1983, Thomas R. Cech and Sidney Altman independently discovered the first known ribozymes, enzymes made of RNA, indicating that the ancient RNA may have been catalytic. However, no RNA molecules that direct the replication of other RNA molecules have been identified in nature. In the mean time, Cech and Jack W. Szostak have modified naturally occurring ribozymes so that they can carry out some of the most important sub-reactions of RNA replication, such as stringing together nucleotides or oligonucleotides.

- In an experiment, Szostak created a pool of random oligonucleotides to approximate the random production presumed to have occurred some 4 billion years ago. From that pool, he was able to isolate a catalyst that could join together oligonucleotides. Also, the catalyst could draw energy for the reaction from a triphosphate group, the very same group that now fuels most biochemical reactions in living systems.
- Another piece of supporting evidence that RNA could have created protein synthesis is that Harry F. Noller, Jr. discovered that it is RNA in ribosomes, not proteins, that catalyses formation of the peptide bonds.
- To explain how self-replicating RNA was created from its constituents, it is hypothesised that the nucleotides in RNA formed when direct chemical reactions led to joining of the sugar ribose with nucleic acid bases and phosphate. Then, these ribonucleotides spontaneously joined to form polymers, at least one of which happened to be capable of engineering its own reproduction.
- First problem with the above hypothesis: in the absence of enzymes, workers have difficulty synthesising ribose in adequate quantity and purity. It has long been known that ribose can be produced easily through a series of reactions between molecules of formaldehyde.

  Yet when such reactions occur, they yield a mixture of sugars in which ribose is always a minor product. The relative paucity of ribose would militate against development of an RNA world, because the other sugars would combine with nucleic acid bases to form products that inhibit RNA synthesis and replication. No one has yet discovered a simple, complete chain of reactions that ends with ribose as the main products.
- ***Second problem :*** Attempts to synthesise nucleotides directly from their components under preiotic conditions have met with only modest success. One series of experiments has yielded purine nucleosides-

units consisting of ribose and a purine base but excluding the phosphate group. Investigators are unable to produce pyrimidine nucleosides efficiently without the aid of enzymes.

- ***Third problem :*** formation of nucleotides by combining phosphate with nucleosides has been achieved by simple pre-biotic reactions, but the kinds of nucleotides that occur in nature arose along with related molecules having incorrect structures. If such mixtures were produced on the young earth, the abnormal nucleotides would have interacted with the normal ones to interfere with catalysts and RNA replication.

  It is thus not easy to see how pre-biotic reactions could have led to the development of the ribonucleotides needed for producing self-replicating RNA. This problem might be solved by the presence of inorganic catalysts or the existence of non-enzymatic reactions leading to efficient synthesis of pure ribonucleotides that scientists have not yet identified.

- ***Fourth problem :*** the condensation of ribose with bases would give complex mixtures of products, including L-as well as D-nucleosides, and nucleosides with alpha-as well as those with beta-glycosidic linkages.

- ***Fifth problem :*** Assuming that ribonucleotides were able to emerge non-enzymatically, we still have to demonstrate that the nucleotides could assemble into polymers and that polymers could replicate without the assistance from proteins. Again, the reactions might be catalysed by minerals. Experiments were conducted to investigate the replication of RNA.

In the experiments, scientists synthesised oligonucleotides and mixed them with free nucleotides. The nucleotides lined up on the oligonucleotides and combined with one another to form new oligonucleotides. Forming such complements from an original template - "copying" - would be the first step in pre-biotic replication of a selected strand of RNA.

Then the strands would have to separate, and a complement of the complemet (a replica of the original strand) would have to be constructed. In practical, scientists were unable to achieve the second step of replication - copying of the complement-without help from protein enzymes. Also, they can induce copying of the original template only when experiments were run with nucleotides having right-handed configuration.

Since on the primitive earth, equal numbers of both kinds of nucleotides would have been present, scientists put equal numbers of both kinds of nucleotides in the reaction mixtures, but copying was inhabited. The problem may be solved if copying without replication have produced a pair of complementary strands and one of the strands happened to be a ribozyme that could copy its complement and thus duplicate itself.

First alternative self-replicating molecule : Eschenmoser proposed a molecule called pyranosol RNA (pRNA) that is closely related to RNA but with an extra carbon atom in the ring. Eschemoser finds that complementary strands of pRNA can combine by standard Watson-Crick pairing to give double-strand units that permit fewer unwanted variations in structure than are possible with normal RNA.

Also, pRNA strands do not twist around each other. In a world without protein enzymes, twisting could prevent the strands from separating cleanly in preparation for replication. The molecule has not been found yet.

- *Second alternative self-replicating molecule:* Peter E. Nielsen used computer-assisted model building to design a polymer that combines a protein-like backbone with nucleic acid bases for side chains. One strand of this polymer or peptide nucleic acid (PNA), can combine stably with a complementary strand. This implies that peptide RNA may be able to serve as a template for the construction of its complement.

Life Originated From Extreme Conditions and its Relation to Minerals

- Earth's microbial ecosystems revealed that in 3.5 billion-year-old sediments from Western Australia appear to have occupied shallow marine hydrothermal environments dominated by island volcanism. In addition, the contemporary microorganisms with the most ancient lineages based on molecular phylogenies are anaerobic, thermophilic, sulphur-metabolising archaebacteria. These organisms were isolated from hot springs and hydrothermal vents where they thrive up to 105°C.
- ***Minerals as containers :*** Some rocks, like gray volcanic pumice, are laced with air pockets created when gases expanded inside the rock while it was still molten. Many common minerals, such as feldspar, develop microscopic pits during weathering. Each tiny chamber in each rock on the early earth could have housed a separate experiment in molecular self-organisation. The hydrothermal origins hypothesis was rejected because amino acids decompose rapidly when they are heated.

Brandes conducted an experiment in which the amino acid leucine broke down within a matter of minutes in pressurised water at 200°C. But when Brandes added to the mix an iron sulphide mineral of the type commonly found in and around hydrothermal vents, the amino acid stayed intact for days.

Minerals as scaffold : Clays are made up of various ions embedded in a two-dimensional silicate lattice. The elements involved are mainly silicon, oxygen, aluminium, iron and magnesium. Clays are formed when water causes the chemical weathering of rocks. The concentration of ions in clays is extremely variable; the surfaces of clays usually have a net negative charge that is neutralised by a positive counter-ion (e.g. Na^+, K^+, Mg^{2+}, Ca^{2+}, Zn^{2+}, Fe^{2+}).

These charges might be able to attract organic molecules

and hold them in place. In the late 1970s an Israeli research group demonstrated that amino acids can concentrate on clay surfaces and then link up into short chains that resemble biological proteins. These chemical reactions occurred when the investigators evaporated a water-based solution containing amino acids from a vessel containing clays.

Separate research teams led by James P. Ferris of the Rensselaer Polytechnic Institute and by Gustaf Arrhenius of the Scripps Institution of Oceanography demonstrated that clays and other layered minerals can attract and assemble a variety of organic molecules. The team at Rensselaer also found that clays can act as scaffolds for the building blocks of RNA.

Minerals as templates : The chemical structure of calcite in many mollusk shells bonds strongly to amino acids. With the help of his colleagues, Robert M. Hazen ran experiments to immerse well-formed, fist-size crystal of calcite into a 50-50 solution of aspartic acid. They observed that calcite's "left-handed" faces selected L-amino acids, and vice versa, with excesses approaching 40 per cent in some cases.

Also, calcite faces with finely terraced surfaces displayed the greatest selectivity. Thus, the researchers speculate that these terraced edges might force the L and D amino acids to line up in neat rows on their respective faces. Under the right environmental conditions, these organised rows of amino acids might chemically join to form protein-like molecules - some entirely of L amino acids, others entirely of D.

Minerals as catalysts : When Brandes and his colleague subjected hydrogen, nitrogen and the iron oxide mineral magnetite to the pressures and temperatures characteristics of a seafloor vent, the mineral catalysed the formation of ammonia, which is required in biological reactions. In 1988 Wachtershauser proposed that minerals - mostly iron and nickel sulphides that abound at deep seafloor vent - could have served as the template, the catalyst and the energy source that drove the formation of biological molecules.

Researchers at Carnegie placed ingredients known to be available to the young earth in capsules and placed the capsules in a massive steel pressure chamber that squeezes the tiny capsules to pressures approaching 2000 atmospheres and heats them to about 250°C. It is found that many common minerals, including most oxides and sulphides of iron, copper and zinc, promote carbon addition by a routine industrial process known as Fischer-Tropsch(F-T) synthesis.

This process can build chainlike organic molecules from CO and hydrogen. It was also demonstrated that F-T reactions can build molecules with 30 or more carbon atoms under some hydrothermal-vent conditions in less than a day.

- ***Minerals as reactants :*** In the catalysis experiments mentioned above, significant quantities of elemental sulphur, organic sulphides, methyl thiol and other sulphur compounds appeared as well. The sulphur in the compounds must have been liberated from the iron sulphide mineral. Liberation of iron was also found. This suggested that iron and sulphur, the elements that form the active centre of certain biological enzymes such as aconitase, can be dissolved from iron sulphide minerals under extreme heat and pressure.

Extraterrestrial Origin of Life

- Scientists suspect that the early days were hot, dry and sterile. It is now clear that space debris bombarded the young planet, creating cataclysms equivalent to the detonation of countless atomic bombs. Impact of this kind, common until 4.0 billions years ago, surely aborted any fledgling life struggling to exist before that time. The short time span for life to emerge implies that the process might have required help from space molecules.
- Since his first report in 1993, John R. Cronin of Arizona State University has demonstrated a slight surplus of left-handed-ness in several amino acids extracted from

two different meteorites. Some believe life's left-handedness is by chance, but extraterrestrial starting ingredients may have predetermined this molecular peculiarity

- Astronomers see signatures of a range of organic compounds throughout the universe, especially among the clouds. For example, a decade of research conducted by Allamandola and others has revealed that polycyclic aromatic hydrocarbons are the most abundant class of carbon-bearing compounds in the universe, trapping as much as 20 per cent of the total galactic carbon in their molecular lattices.
- Experiments revealed that even at the extremely low temperatures and pressure of space, the UV radiation can break chemical bonds. When the atoms are locked in ice, this bond-breaking process can make molecular fragments re-combine into unusually complex structures that would not be possible if these fragments were free to drift apart.

Bertein started with a simple ice of frozen water, methanol and ammonia - in the same proportions seen in space ice - the experiment yielded complex compounds such as the ketones, nitriles, ethers and alcohols found in carbon-rich meteorites. They also created hexamethylenetetramine, or HMT, a six-carbon molecule known to produce amino acids in warm, acidic water.

David W. Deamer found that some of the molecules in the cloud-chamber ice grains form capsulelike droplets in water. These capsules are strikingly similar to extracts of meteorites from Murchison.

Researchers found that interstellat amorphous ice when exposed to radiation such as that found in deep space, it too can flow. Thus, it could be an explanation of how organic molecules may endure and react within the ice.

Emerging consensus in planetary science agree that the

early pre-biotic atmosphere is a neutral one rich in carbon dioxide and molecular nitrogen. Early CO_2-rich atmospheres are implied by "hot" accretion scenarios for Earth, in which core formation takes place quickly, leaving the upper mantle in an oxidised state.

The short photodissociation lifetimes of CH_4 and NH_3 in model paleoatmospheres reinforce here. There is a dense, 10- to 20-bar CO_2 early terrestrial atmosphere, consistent with the early faint sun "paradox". Synthesis of key pre-biotic molecules such as hydrogen cyanide and formaldehyde would have been much more difficult in CO_2 atmospheres than in reducing ones.

A long standing objection to extraterrestrial origin is that the organic compounds would be totally dissociated by the heat of cemetery atmosphere passage and the ensuing impact. However, researchers speculated that aero-braking (slowing by atmospheric drag) and uneven distribution of shock energy throughout the impacting projectile will conspire to yield some region of the comet for which temperatures remain low enough to allow at least the hardier organics to survive.

Because gas-phase results on shock pyrolysis are not available, it is estimated that alanine could withstand temperatures of ~700K for 1s, whereas other amino acids should withstand temperatures in the range of 600 to 800K. Through modelling, it is shown that dense CO_2 atmospheres allow intact cometary organics to be delivered in large amounts.

7

Temperature Regulation

The word 'temperature' may be defined as the degree of hotness or coolness of any thing, or, specifically as the degree of heat of a living body, usually as measured by a thermometer.

However, active life is limited to a narrow temperature range from about -2°C to about + 50°C. Beyond this range many animals can survive in an inactive or torpid state. A large number of animals, including all aquatic invertebrates, have the same temperature as their surroundings. But birds and mammals generally maintain a nearly constant temperature in spite of wide variations in the surrounding temperature. Thus, animals are of the following types:

Cold-blooded (Poikilothermic) Animals: A 'cold-blooded' animal is not necessarily cold, a desert lizard or an insect sitting in the sun may have a higher body temperature than a mammal; but it has nearly the same temperature as its surroundings. For example, a fish has the temperature of the water it swims in, and an earthworm has the temperature of the soil it lives in. However, the corresponding term 'poikilothermic' (Greek *poikilos* = changeable) is really more descriptive, for it signifies changeable temperature. In other words, it refers to the fact that the temperature of the cold-blooded animal fluctuates with that of their surroundings.

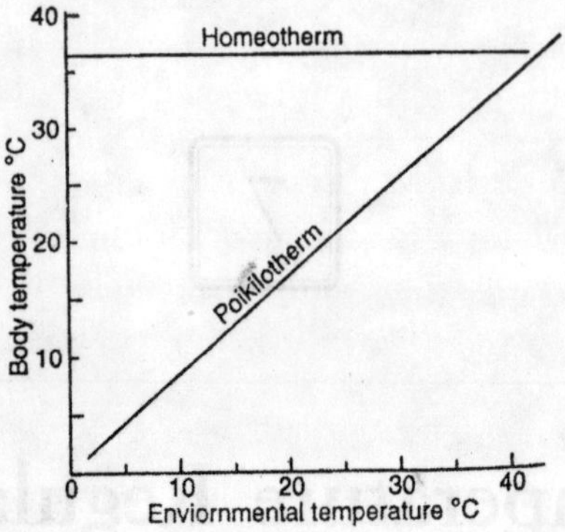

Warm-blooded (Homeothermic) Animals: A 'warm-blooded' animal is not necessarily always warm. A few mammals and some birds undergo hibernation during which their temperature may decrease to near the freezing point of water, with no harm to the animal and in that state it seems odd to call them warm-blooded.

The corresponding term 'homeothermic' (Greek *homos* = like, and *homoios* = similar) or homeothermic or homoiothermic is a better expression than 'warm-blooded' for it signifies the constant temperature. In other words, mammals and birds are called homeothermic because they maintain a quite constant body temperature in spite of wide variation in the surrounding temperature. Thus, a polar bear, a tropical monkey and a walrus have almost the same body temperature of about 38°C.

Heterothermic Animals: The term 'heterothermic' (Greek *heteros* = different) is used to refer to the animals that at times have a high and well-regulated body temperature, but at other times are more like the cold-blooded animals.

Ectothermic and Endothermic Animals: By exposing itself to the sun, a lizard can maintain its body temperature far above that of the surrounding air. Hence, to distinguish the lizard from the birds and mammals, which maintain their high body temperature by metabolic heat production, we use the term ectothermic and endothermic. *Ectothermic* animals are those which depend on external heat sources, primarily the solar radiation, to maintain their body temperature. *Endothermic* animals are those which are able to maintain a high body temperature by internal heat production.

However, these terms and definitions also have their limitations. For example, a tuna fish may keep the temperature of its muscles 10 to 15°C above that of the water. The heat is derived from muscular metabolism, but we do not call the tuna an endothermic animal in the same sense as the birds and mammals.

Temperature Regulation in Poikilothermic Animals: Although the temperature of a poikilothermic animal depends on that of the surroundings, this does not mean that the animal has no control whatever over its body temperature. Some poikilothermic animals can and indeed do stay warmer than the air or water in which they live. The heat balance equation for a poikilothermic animal may be written as follows:

$H_s = H_{tot} \pm H_r$ — He—H_e where

H_s = storage of heat in the body

H_{tot} = total heat production

H_r = heat exchange by radiation

H_c = heat exchange by conduction and convection

H_e = heat exchange by evaporation

If the body temperature is higher than the surrounding medium, the terms for conduction (H_e) and evaporation (H_e) will signify heat losses and be negative. Hence, if the animal has to increase the body temperature, it has to do something to maximise the heat storage (H_s). The mechanisms vary in different groups of poikilotherms and are as follows:

Aquatic Poikilothermic Animals

For an aquatic animal the heat balance equation may be written as follows:

$$H_s = H_{tot} — H_e$$

because in water there is no heat exchange by evaporation and radiation. In order to increase heat storage, either the total heat production is to be increased, or the heat loss by conduction is to be reduced. Because of high thermal conductivity and high heat capacity of water, small animals lose heat rapidly and are unable to attain a body temperature very different from the medium.

However, some animals are able to attain a high body temperature with the help of a *'heat exchanger'* placed between the gills and tissues. For examples, the tuna and some sharks maintain a high temperature in their swimming muscles, independently of the water in which they swim. In the ordinary fish the swimming muscles are supplied with blood from the dorsal aorta. But in the tuna the muscles used for steady, fast swimming (the so-called dark red muscles) receive their supply from the blood vessels running along the side of the fish just under the skin. From these major vessels arise a large number of parallel fine blood vessels which form a slab in which arteries are densely interspersed between veins running in the opposite direction. This puts the cold end of the heat exchanger at the surface of the fish and the warm end deep in the muscles.

The arterial blood from the gills is at water temperature, and as this blood runs in the fine arteries between the veins, it picks up heat from the venous blood which comes from the muscles. The exchange of heat is facilitated by the small diameter (about 0.1 mm) of the vessels. When the venous blood has reached the larger veins under the skin, it has lost its heat, which is returned to the muscle via the arterial blood. Consequently, the tuna can maintain muscle temperatures as much as 14°C higher than of the water in which it swims.

Thus, with the help of *heat exchangers* the swimming muscles are kept warm. The advantage is that the high temperature increases the power output of the muscles. At higher temperature the muscle contracts faster, and since the number of contractions per unit time increases, the work per unit time (power output) increases accordingly. A high power output gives the tuna a high swimming speed relatively independent of the water temperature, and this in turn makes it possible to pursue successfully the prey which otherwise swims too fast to be caught, such as the squid or the mackerel. In the tuna, separate heat exchangers permit the digestive organs and the liver to be kept at a higher temperature. This increases the rate of digestion and in turn the rate of fuel supply to the muscles for their high power output.

Fish living in cold arctic waters appear to have an *anti-freeze substance* in their tissues that prevents freesing when temperatures fall below those at which their body fluids would normally freeze.

Terrestrial Poikilothermic Animals and Insects: In order to raise the body temperature in terrestrial animals the evaporation and heat loss by conduction are to be reduced, and the heat gain by radiation and metabolic heat production are to be increased. The only commonly available radiation source is the sun, and in the absence of solar radiation, the metabolic heat production is the only way to increase the heat gain. Solar radiation is used especially by the insects and reptiles. To increase the amount of radiant energy absorbed, they depend both on their colour and on their orientation relative to the sun. Many reptiles can change their colour by dispersion or contraction of the black pigments in their skin. Since about half of the solar radiation energy is in the visible light, a dark skin increases the amount of solar energy that is absorbed rather than reflected.

The other way of increasing the heat gain from solar radiation is to increase the exposed area. This is done by orienting the body at right angles to the direction of the sun rays, and by flattening the body and spreading the legs. In this way lizards can attain the temperature much higher than the surrounding air. When a suitable temperature has been attained, further heating is avoided by lightening the skin colour and changing the posture to avoid direct sun rays.

Lizards and other reptiles burrow at night when the desert or mountain temperatures are at their lowest. Some heat from the day remains underground and the reptile's body is prevented from freezing. At sun rise the lizard, which may have reached a body temperature as low as 5°C during the night slowly emerges from its burrow to orient itself to the sun light so that the maximum heat is obtained. By proper orientation lizards can attain body temperatures that are as much as 30°C above that of the air.

In terrestrial poikilotherms, internal heat production can also be used to attain a high body temperature. It has been reported that some female snakes do this while incubating their eggs and it is done by increasing the rate of oxygen consumption. The increase in oxygen consumption seemed to be caused by strong muscular contractions reminiscent of shivering in mammals.

Insects have a special thermal problem because of their high rate of metabolism during the flight. An insect can increase the temperature of the flight muscles by a process similar to shivering in man. Such heating of the muscles before take-off occurs mainly in large insects such as the locusts, large moths, butterflies, and bumblebees, and also in the wasps and bees which can fly rapidly. *Heinrich* and *Bartholomew* (1971) have made an analysis of the pre-flight warm-up in the sphinx moth, *Manduca sexta*. During the warm-up the animal seems to shiver as the wings vibrate lightly. The heat is generated by the contraction of muscles, and the reason that the wings do not flap is that the muscles for the upstroke and downstroke contract simultaneously, instead of alternately as in the flight.

Honeybees regulate not only their own individual temperature, but that of the entire hive as well. In the summer, the temperature of a beehive is maintained at about 35°C and fluctuates very little. This provides the colony with an optimum temperature for the development of the brood.

If the beehive temperature rises above the regulated level, the bees spread droplets of water over the comb, and evaporation is aided by air currents created by a large number of bees fanning their wings. Below 30°C honeybees *Apis mellifera* do not usually consume any water, but at higher temperatures their water intake increases considerably as the water is used for cooling. (*Tree* and *Spencer-Booth,* 1950).

The *clustering* of bees is a heat conservation mechanism, and bees reflexively cluster upon stimulations involving the sight, sound, odours of abdominal scent glands and the temperature. The minimum number of bees that will cluster depends upon the temperature. In fact, there is a tendency for the cluster temperature to be higher at the very lowest ambient temperatures. For each 5°C fall in the outside air temperature, the cluster temperature increases by about 1°C.

At lower temperatures a smaller number of bees can be stimulated into clustering than at higher temperatures, where perhaps 50 bees are a minimum number that can be induced to cluster. This thermal regulation of the hive, which is needed for proper reproduction and for the development of young bees, is based on both physiological heat production and behavioural mechanisms associated with clustering and wing fanning. Bees may also cool a hive during warm weather by standing at the hive entrance and fanning their wings to produce a current of air through the hive. Ants and other social insects exhibit similar behaviour in maintaining colonies at proper temperatures.

Temperature Regulation in Homeothermic Animals: For a body to maintain a constant temperature, the heat loss must be equal to the heat gain. The "Du Bois temperature balance" shows graphically how the physiological and metabolic reactions which produce heat must be matched against those which radiate or conduct it away in order to provide a constant body temperature.

Heat Loss: It occurs by the following means.

(a) *Conduction:* Conduction of heat consists of direct transfer of the kinetic energy of molecular motion, and it always occurs from a region of higher temperature to one of the lower temperature. The loss of heat by ***conduction to air*** represents a sizable proportion of the body's heat loss (about 15 per cent).

(b) *Convection:* The transfer of heat in fluids is almost invariably accelerated by the process of convection, which refers to mass movement of the fluid. The free or *natural convection* is caused by temperature difference. Convection caused by the external forces, such as the wind, water currents, etc., is referred to as the *forced convection.*

(c) *Radiation:* Heat transfer by Radiation takes place in the absence of direct contact between the objects. All objects at a temperature above absolute zero emit electromagnetic radiation. The intensity and the wavelength of this radiation depend upon the temperature of the radiating surface and its emissivity. The human body radiates heat in all directions.

(d) *Evaporation:* The amount of heat required to achieve the phase change from liquid water to vapour is known as the *heat of vaporisation (Hv)*. The heat of vaporisation changes slightly with the temperature at which the evaporation takes place. Thus, at 0°C the Hv is 595 cal/gram water, at 22°C it is 584 cal/gram, and at 100°C it is 539 cal/gram. In physiology, it is customary to use the figure 580 cal/gram, which is an approximation of the value for vaporisation of water at the skin temperature of a sweating man, about 35°C. Heat loss due to evaporation takes place as follows.

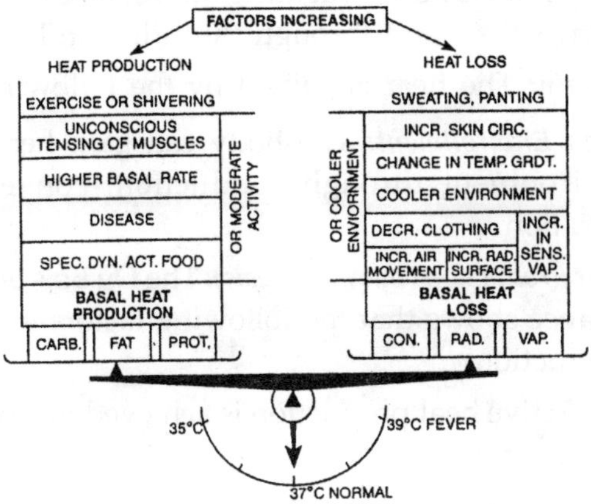

The Du Bois temperature balance (balance between the factors increasing heat production and heat loss).

(i) *Active Heat Loss:* When a man is exposed to hot surroundings he cools himself by evaporation of sweat from the general body surface. *(sweating).* Stimulation of the anterior hypothalamus (preoptic area) by excess heat causes sweating. The impulses from this area are transmitted in the autonomic pathways to the cord and then through the sympathetic outflow to the skin. The sweat glands are innervated by sympathetic *cholinargic* nerve fibres.

However, a dog evaporates most of the water from the respiratory tract by panting, a very rapid, shallow breathing which increases evaporation on by the thermoregulatory centres of the brain (hypothalamus). The actual panting process is controlled by a panting centre that is closely related to the pneumotaxic respiratory centre in the pons.

Some animals use a third method for increasing evaporation. They spread saliva over their fur and lick their limbs, thus achieving cooling by evaporation.

(ii) *Basal Heat Loss:* This includes evaporation of water from the lungs during respiration, insensible perspiration, warming of respired air in cold weather, and through the urine and faeces.

Heat Gain: The heat is gained by the following means.

(i) *From Environment:* The body can gain heat from the environment through conduction, convection and radiation.

(ii) *Heat Production (thermogenesis):* The Du Bois temperature balance shows that the following factors increase heat production.

(a) Active heat production is achieved through *exercise* and *shivering*.

(b) Basal heat production depends on (i) unconscious tensing of muscles, (ii) disease, and (iii) specific dynamic action of food. The extent to which a particular food item influences the rate of metabolism is referred to as its specific dynamic action. Proteins have a higher specific dynamic action. As they raise the metabolic rate by about 30 per cent, they help in warming up the body in cold weather.

In homeothermic animals the thermogenesis depends on the following:

1. *Shivering Thermogenesis:* A man, when appropriately acclimated, may show an eighteen-fold increase in heat production within 12 minutes if suddenly immersed in a bath at 4°C. This record heat production is associated with the shivering or the *involuntary muscle contraction.*

Shivering can be evoked either by a slight decrease in the core temperature or by a decrease in the skin temperature. Thus, its control can be either central or peripheral or both. The conspicuousness and importance of shivering varies from animal to animal. Among mammals, the most obvious shiverers are probably the fruit bats or flying foxes of the Old World. Some flying foxes of the genus *Pteropus* shiver strongly and visibly at all temperatures below their thermal neutral zone.

2. *Nonshivering Thermogenesis:* The heat produced from extramuscular sources is termed non-shivering thermogenesis. During cold stress, acclimated rats produce as much as 65 per cent of their heat in fat and visceral organs.

Nonshivering thermogenesis plays an important role in the acclimation of mammals to the low temperatures and in arousal from the hibernation and daily torpor. It is clear that the capacity for non-shivering thermogenesis acquired during acclimation to cold allows rats to be active at much lower temperatures than would be the case if they had to depend on shivering alone, for exercise eliminates shivering but does not affect the non-shivering thermogenesis. Exercise and non-shivering thermogenesis are additive, exercise and shivering are not. Non-shivering thermogenesis also significantly increases metabolic scope for activity. In cold-acclimated rats with a high level of non-shivering thermogenesis, the metabolic scope below 30°C is almost the twice that of the warm-acclimated rats at the same temperature.

The increase in non-shivering thermogenesis associated with the cold adaptation in animals involves shifts in the pathways and relationships of the intermediary metabolism. Although the situation is not completely understood, it seems probable that the cold-acclimated rats have increased capacity for the formation and utilisation of glucose.

It also seems clear that some non-shivering calorigenesis is evoked in part through the hormonal stimulation of lipid metabolism. For example, the colorigenic response to injection of norepinephrine in cold-acclimated rats involves a marked increase in oxygen consumption and a rise in body temperature. Both these responses can be related to in increased capacity of the liver to form and mobilise non-esterified fatty acids. The thyroid hormone plays an important role in the long term development of mammalian non-shivering thermogenesis. One of the important sites of shivering thermogenesis is the brown fat (see below).

3. *Special Thermogenic Tissue:* Many mammals have a special thermogenic tissue known as the *brown fat* or the brown adipose tissue. Brown fat was first described in hibernating marmots, *Muris alpinus*. It was called the hibernating gland and assumed to play an active role in hibernation. The tissue is localised in the neck and thoracic regions in relation to the major blood vessels so that its heat is quickly transported to those organs (heart and brain) whose continuous high temperatures are vital for a homeotherm. Histologically, the brown fat is an extremely vascular tissue with relatively high concentrations of myoglobin, cytochromes, and flavin compounds. These give it a pale buff to dark reddish brown colour which contrasts with the ordinary white adipose tissue. It also differs cytologically from the white fat in that the cells are smaller, polygonal, with central nuclei and numerous fat droplets which are dispersed rather than concentrated in a single vacuole; mitochondria are small, numerous and closely associated with the fat droplets. Brown fat also differs biochemically from the white fat, with higher concentrations of water, protein, phospholipid, cholesterol, and certain mitochondrial enzymes. Details of mitochondrial respiration in the brown fat are speculative but heat, rather than ATP, appears to be the main end product. This tissue can be thought of as a rapidly firing furnace that warms up some mammals quickly in emergency, without muscular activity or shivering.

The brown fat is most conspicuously developed in many newborn mammals (including the human), in hibernating animals at the time of arousal, and in cold-acclimated animals.

Like a smoothly operating automatic furnace, it responds to cold stress with a large burst of heat. The physiological mediator for the stimulation of brown fat appears to be the noradrenaline. Sympathetic stimulation activates the adenyl cyclase system with rapid lipase hydrolysis and oxidation of fats.

Heat Balance: Body temperature is controlled by balancing heat production against heat loss. After considering the means of heat loss and heat gain, the heat balance equation for a homeothermic animal may be written as follows:

$$H_{tot} = \pm Hc \pm Hr \pm He \pm Hs$$

where H_{tot} = metabolic heat production (always positive)

H_c = conductive and convective heat exchange

H_r = net radiation heat exchange

H_e = heat loss in evaporation

H_s = storage of heat in the body

The three components of heat exchange, conduction, radiation and evaporation depend on the external factors ; of them the most important single factor is the temperature. It is obvious that the heat losses increase when the external temperature falls. If, on the other hand, external temperature rises, the heat losses decrease, and if external temperature exceeds the body surface temperature, heat exchange by both conduction and radiation may be from the environment to the organism. The total heat gain is then the sum of heat gain from metabolism plus heat gain from the environment. This situation still permits the maintenance of a constant body temperature, provided that evaporation is increased sufficiently to dissipate the entire heat load.

Cold and its Application

Temperature regulation in cold is for keeping the body warm. To maintain constant body temperature we must satisfy the steady state condition in which the rate of metabolic heat production (H) is equal to the rate of heat loss (Q). In cold the heat loss takes place only through conduction (including convection) and through radiation, and that heat loss through evaporation can be disregarded. The following equation relates heat loss to ambient temperature:

$$H = Q = C(T_b - T_a)$$

It says that the rate of metabolic heat production (H) equals the rate of heat loss (Q), which in turn is proportional to the temperature difference seen the environment (T_b-T_a) The term C is a conductance term. Among the terms in the above equation there is very little an animal can do about the ambient temperature (T_a). Hence, to adjust to an unfavourable T_a an animal has only the three remaining terms, i.e., the heat production (H), the conductance term (C), and the body temperature (T_b) that could possibly be adjusted. Since we are concerned with the maintenance of the body temperature (T_b), we are left with adjustments to be made either in the heat production (H), or in the conductance term

Increase in the heat production is achieved by (i) muscular activity and exercise, (ii) shivering, and (iii) non-shivering thermogenesis. Conservation of heat is brought about *by piloerection peripheral vasoconstriction* and *countercurrent* warming of blood. Conductance, in the sense it has been used above, is a measure of heat flow from the animal to the surroundings. It includes the flow of heat from deeper of the body to the skin surface and from the skin surface through the fur to the environment. When conductance is low, it means that the insulation value is high. The insulation values of the fur from various animals differ considerably.

The insulation value increases with the thickness of the fur and is maximum in the animals which have a thick fur, such as the white fox. Among the smàller animals there is a clear correlation between the thickness (and insulation) and the size of the animal. They must have relatively short and light fur or the animal could not move about. This is particularly true for small rodents and shrews, the smallest mammals. Because of their relatively poor insulation, they either take advantage of microclimates (*e.g.*, by living in burrows), or escape from the problem of keeping warm by hibernating. It is also well known that the thickness of the fur changes with the seasons, and that the winter fur is heavier and presumably affords better insulation than the thinner summer fur.

In arctic land mammals the temperature gradients are different, the surface temperature of the body skin under the fur is regularly within a few degrees of the core temperature (*Irving* and *Krog*, 1955) and most of the insulation, therefore, resides outside the skin surface. The polar bear, which has a fur that loses most of its insulation value in water, also has a substantial layer of blubber without which the semi-aquatic way of life would be impossible. The main difference in the insulation between aquatic and terrestrial mammals is that the insulator of the aquatic mammal (the blubber) is located internally to the surface of heat dissipation. Therefore the blood can bypass the insulator, and heat dissipation during heavy exertion or in warm water can be independent of the insulator. In contrast, terrestrial animals have the insulator located externally to the skin surface. Hence they cannot modulate the heat loss from the general skin surface to any great extent.

Conservation of Heat Energy

It is achieved by the following methods:

Pilo-erection: It means the erection of hair. In hot weather the hair of homeothermic animals lie flat; but in cold, hair are erected and the animal appears more or less coated with fur. The latter acts as an insulator and thus heat loss is decreased. The erection of hair is achieved by small muscles at the root of the hair and is controlled by the central nervous system in response to the information received from the thermal receptors of the skin. Thus, pilo-erection is an important physiological response to cold in all homeotherms, including man.

Peripheral Vasoconstriction: Conservation of heat is also achieved by peripheral vasoconstriction, *i.e.,* the constriction of skin blood vessels. It reduces the amount of blood flowing through the skin and, therefore, the heat loss from the skin surface is reduced. Vasoconstriction usually takes place in the covered parts of the body.

Countercurrent Heat Exchanger: The countercurrent heat exchanger is based on a particular arrangement of the circulatory vessels in which the blood flows in the opposite direction. In the peripheral regions of the body, for example, the arterial blood at the higher core temperature flows to a region of lower temperature.

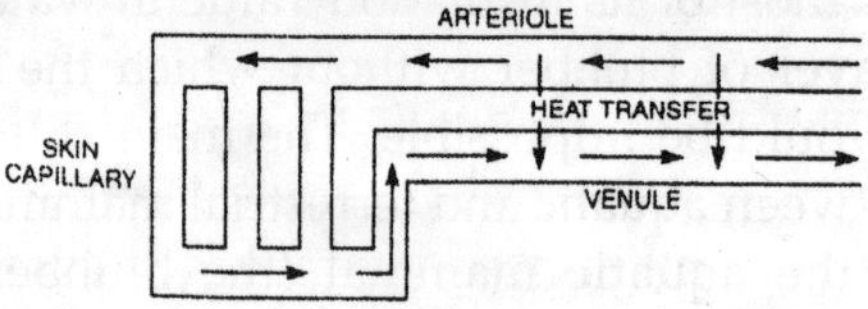

Diagrammatic representation of the thermal countercurrent exchange system.

The venous blood that has passed through the peripheral region is at a lower temperature than the arterial blood because of heat loss to the environment. Because of the contact of the vein with the artery, the heat of the arterial blood is passed to the returning venous blood. Through such conduction of heat from the artery to the vein, the blood that reaches the periphery does not lose as much heat to the environment as would arterial blood that had not transferred some heat to the venous blood. Such a system provides for an exchange of heat to core-directed blood but still permits the blood to carry essential nutrients and other materials to peripheral cells. The system also ensures that the venous blood will not have a temperature low enough to act as a stimulus for the thermogenic machinery. The countercurrent heat exchangers are present in many regions of the mammalian body. They are found, for example, in the circulation of the digestive tract and are needed there so that the cold liquids or foods entering the digestive system do not remove too much heat from the lining of the tract. The countercurrent exchange systems also operate in the brown adipose tissue. In the flippers and flukes of whales and seals, major arteries are completely surrounded by veins, allowing for a maximal heat exchange between the arterial and venous bloods. This exchange prevents too much body heat being lost from the broad surfaces of their flippers and flukes to the cold marine waters in which these animals live. When it is necessary for heat to be lost, blood is shunted to other venous pathways that are not in contact with arteries. In this way the circulatory system of the flipper can function both in heat conservation and in heat dissipation.

Although thermal countercurrent exchange systems are especially prominent in the animals that live in colder environments, such arrangements are found in the animals of both the temperate and tropical environments as well. For example, the sea cows, which live in tropical and subtropical water, have similar heat exchangers in their appendages. Heat exchangers in the limbs are not restricted to aquatic animals. Even in the limbs of man some heat exchange takes place between the main arteries and the adjacent larger veins located deep within the tissues. In birds, heat exchange in the legs is very important, especially for the birds that stand or swim in water. Unless the blood flowing to these thin-skinned peripheral surfaces went through a heat exchanger, the heat loss would be very great indeed. A gull placed with its feet in ice-water for two hours lost only 1.5 per cent of its metabolic heat produced from the feet (*Scholander*, 1955).

Thus, due to the heat exchanger the heat loss is quite insignificant. Some tropical animals like sloth also have heat exchangers in their arms.

Heat and its Applications

It has been described that the birds and mammals increase their heat production in order to stay warm at the ambient temperature below a certain point, the lower critical temperature. Above the critical temperature the metabolic heat production remains constant through a range usually known as the thermoneutral range. So, for the body temperature to remain constant, metabolic heat must be lost at the rate it is produced. In the heat balance equation:

$$H = C(T_b - T_a)$$

If H and T_b, remain constant and T_a changes, the conductance term (C) must also change. The conductance term refers to the total heat flow from the organism, and it can be increased in several ways.

In heat regulation in cold the role of evaporation was, however, disregarded because it made up a fairly small fraction of the total heat exchange. But in hot climate (at higher temperature) evaporation is the key factor in the heat balance. When some mammals and birds are kept in air that equals their body temperature, those that are able to keep their body temperature from rising evaporate an amount of water which is equivalent to their metabolic heat production. When the ambient temperature exceeds the body temperature, as in the desert, the evaporation is used to dissipate the sum of metabolic heat production and the heat gain from the environment.

In humans, *sweating* is the means for evaporation and they sweat to increase cooling by evaporation. However, several furred animals, including cattle, large antelopes and camel also depend mainly on sweating for evaporation. As dogs have a few sweat glands, they cool primarily by *panting*, a very rapid shallow breathing which increases evaporation from the upper respiratory tract. A number of smaller ungulates, goats, sheep, small gazelles and many carnivores also depend on panting for evaporation. In some animals, for example, marsupials, a third method is used to increase evaporation. They spread saliva over their fur and lick their limbs, thus achieving cooling by evaporation. Birds, in the absence of sweat glands, increase evaporation either by panting or by a rapid oscillation of the thin floor of the mouth and upper part of the throat, a mechanism known as the *gular flutter.* They may use one or both the mechanisms at a time (*Lasiewski* and *Crawford,* 1968).

Some animals, such as the African gazelles, and domestic ungulates, etc., have the mechanism for selective cooling of blood to the brain. Some reptiles, when exposed to heat, increase their respiration frequency and breathe with an open mouth, a situation reminiscent of panting.

Besides, some animals are able to maintain different parts of the body at different, well-regulated temperatures. For example, in the desert lizard, *Sauromalus obesus,* the carotid arteries run very close to the surface of the pharynx. As the arteries pass right under the moist surfaces where evaporation takes place, the blood is cooled before it enters the brain. Similarly as mentioned above, in the gazelles and ungulates the brain is supplied with cool blood.

In these animals most of the blood to the brain flows in the external carotid artery, which at the base of the skull divides into hundreds of small arteries, which then rejoin before passing into the brain. These small arteries lie in a large sinus of venous blood which comes from the walls of the nasal passages where it has been cooled. The blood that flows through these small arteries is, therefore, cooled before it enters the skull, and as a result the brain temperature may be 2 or 3 degrees lower than the blood in the carotid artery and the body core.

Relations between Body Size and Heat Regulation

In hot conditions the total heat gain consists of: heat gain from the environment, and metabolic heat gain. The heat gain from the environment by conduction and convection, and radiation from the sun and from the ground, are surface processes, and the total environmental heat load, therefore, is directly related to the surface area. Therefore, the total heat load or the sum of environmental and metabolic heat gain, is roughly proportional to the body surface area.

According to *Schmidt-Nielsen* (1964), for a mammal to maintain a constant body temperature under desert conditions, water must be evaporated in proportion to the heat load. Because of the larger relative surface area of small animals, the heat load, and therefore the estimated evaporation in relation to the body size, increases rapidly in the small animal. For example, if a man in the desert evaporates at a rate of one litre per hour (which equals about 0.60 litre per metre^2 body surface per hour), small rodents weighing between 10 to 100 g will have to evaporate at a rate of 15 to 30 per cent of their body weight per hour . Since a water loss of 10 to 20 per cent is fatal for mammals, such water loss is impossible.

Hence, desert rodents, because of their small size alone, must escape from heat, which they do by going inside their underground burrows during the day. However, the small squirrels of the North American deserts go on moving in and out of their cool burrows in the daytime. By doing so they can rapidly pick up heat and unload it again and again without using water for evaporation. This permits them to be active during the day.

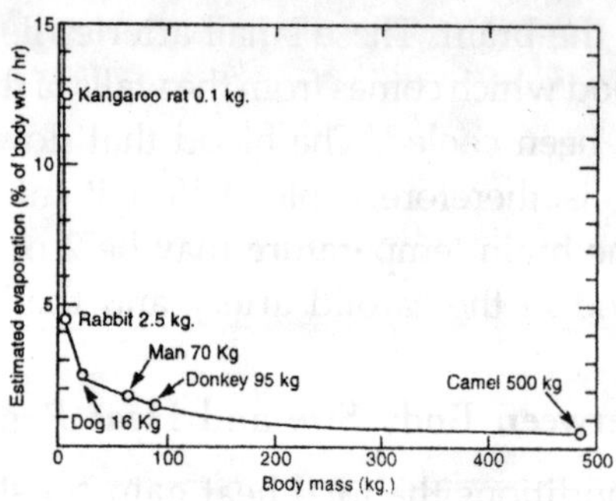

Estimated evaporation in relation to the body size. The curve is calculated on the assumption that heat load is proportional to body surface.

A large animal, such as a camel, evaporates less water and derives a considerable advantage simply from being big. As the camel uses several methods to reduce heat gain, a lower-than-predicted (from body size alone) water loss is due to a number of reasons: (a) heat is stored because of an increase in body temperature (in a camel deprived of water the morning temperature may be as low as 34°C and the late afternoon temperature may be about 41 °C, this 7°C increase in body temperature corresponds about 29000 kcal of heat which equals a saving of 5 litres of water), (b) the increase in body temperature decreases the heat flow from the environment, (c) the fur reduces the heat gain from the environment, and (d) it can tolerate a greater degree of water depletion of the body, and when water is available, it may drink more than one third of its body weight.

Role of Hypothalamus in Temperature Control in Humans: The regulation of body temperature is very precise and is controlled by the *temperature-regulating centres* located in the *hypothalamus*. It has been found that the anterior hypothalamus contains heat-sensitive neurons and the posterior hypothalamus contains cold-sensitive neurons. The heat-sensitive neurons increase their firing rate as the temperature rises. The cold-sensitive neurons, on the other hand, increase their firing rate when the body temperature falls.

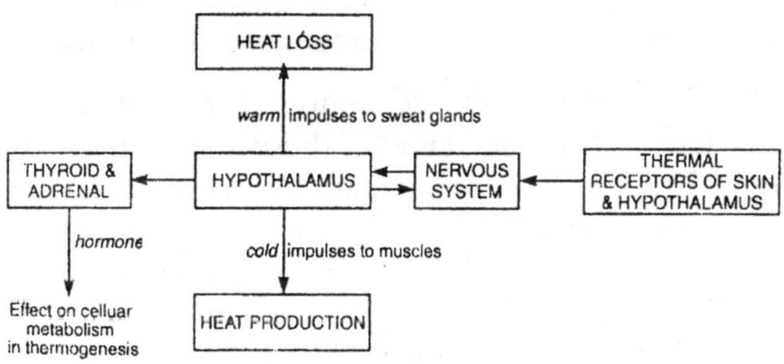

Diagrammatic representation of the role of hypothalamus in temperature regulation. The hypothalamus receives information about the temperature of body surface via afferents from thermal receptors in the skin, and it also detects changes in the temperature of the blood by means of receptors in the hypothalamus itself.

When the body needs more heat, the hypothalamus sends nerve impulses to the muscles, causing contractions and shivering. On the other hand, if the body needs cooling, impulses from the hypothalamus to the sweat glands induce sweat production. The temperature-regulating centre, therefore, acts like a thermostat. Thus, hypothalamus has two roles in thermoregulation. It integrates all of the sensory information available to the nervous system about internal and external temperature conditions and also acts by itself as a sensory mechanism for determining the core temperature. The hypothalamus also acts in integrating homeostatic functioning. Temperature regulation, for example, cannot cause such losses of water by sweating or panting that would disturb the osmoregulation.

The hypothalamus exerts its effects in temperature regulation not only by causing the output of impulses along various motor neurons, as mentioned above, but also affects the thyroid and adrenal glands whose hormonal secretions can influence the cellular metabolism in thermogenesis. The increased thyroxine output by the thyroid gland increases the rate of cellular metabolism throughout the body, which is yet another machanism of *chemical thermogenesis*. Thus, the integrating role of hypothalamus in temperature regulation is further complicated by its neuroendocrine activities.

Any disturbance in the temperature-regulating centre results in fever *(pyrexia)*. It is a condition in which a higher than normal temperature is maintained by the hypothalamus. It is *pyrogens* (fever-producing substances) released grain toxic bacteria or pyrogens released from degenerating tissues of the body that cause fever during disease conditions.

The onset of fever is marked by all the reactions normally seen when a man is exposed to cold. There is shivering, pilo-erection, cutaneous vasoconstriction, and an increase in thermogenesis. The rate of heat loss is decreased, the rate of heat production rises and thus the core temperature is increased. On the other hand, termination of fever is marked by those reactions normally seen upon exposure to high temperature as the body cools off. Drugs such as aspirin that reduce the level of fever are called anti-pyretics. They do not reduce normal temperature, but they probably act by antagonising the actions of *pyrogens* on the hypothalamus. They reset the hypothalamic thermostat and the temperature returns towards normal as a result of increased heat loss brought about by peripheral vasodilation and sweating.

Behavioural and Adaptive Temperature Controlling Mechanism: Besides the physical and chemical mechanisms behavioural mechanisms are used by a variety of organisms, including poikilotherms, to at least periodically maintain a body temperature higher than that of the environment.

In humans, *behavioural control of body temperature* is yet another temperature-controlling mechanism that is even more important. Whenever the internal body temperature becomes too high, signals from the hypothalamus give the person a psychic sensation of being overheated. Conversely, whenever the body becomes too cold, signals from the skin and other receptors elicit the feeling of cold discomfort. Therefore, the person makes appropriate environmental adjustments to regain comfort, such as moving into a heated room in freezing weather. This is the only really effective mechanism for body heat control in extremely cold environs.

Some behavioural adaptations are especially interesting because they extend the resources of the organism beyond the range of its normal physiological reactions. They are as follows:

Hibernation: It is difficult to give a completely satisfactory definition of the term hibernation. The word hibernation (from Latin *hiberna*- winter) is sometimes used to designate an over-wintering, inactive stage of poikilothermic animals such as the insects and snails, etc. The hibernation of homeothermic animals may be defined as a state or a torpid condition in which the body temperature is greatly decreased and the metabolism, respiration, heart rate, etc., are considerably reduced. Thus, the state of hibernation allows the animal to remain dormant during the winter when the food is less readily available.

Mammals that hibernate belong to three orders—insectivora (e.g., the European hedgehog *Erinaceous europaeus),* rodentia (e.g., the golden hamster *Mesocricetus auratus,* the European dormouse *Glis glis,* and the little pocket mouse *Perognathus longimembris,* the American woodchuck *Marmota monax,* the European marmot *Marmota marmota,* and a variety of ground squirrels including *Citellus beecheye. C. franklini, C. lateralis,* etc.), and chiroptera *(e.g.,* the little brown bat *Myotis lucifugus).* Hibernators are also found among the Australian marsupials, such as the pigmy opossum. Hibernating birds include hummingbirds, the insect eating swifts, and some mouse birds *(e.g.,* the African genus *Colis).*

Hibernation is a phenomenon that is under accurate physiological control. The hibernator is in a state comparable to that of a poikilotherm and its body temperature is close to that of the surroundings and rises and falls with it. When the temperature increases appreciably, the animal warms up again and returns to the homeothermic state. During hibernation the oxygen consumption may decrease to 1/30 to 1/100 of the normal, the heart rate goes down to a few beats-per minute, and the respiratory movements become very slow. It may appear that the major change in a hibernator is the failure of the temperature-regulating centre. But this is not the case because it has been found in some hibernators that if the temperature drops to below freezing, heat production in the animal increases to prevent its tissues from freezing. Thus, the hibernator does not behave as a poikilotherm and hibernation is a well regulated physiological state that permits survival during the colder seasons when the food is in short supply.

However, bears are not considered true hibernators, for although they sleep during the major part of the winter, their body temperature does not decrease more than a few degrees and they show only a moderate drop in metabolic rate and other physiological functions. Bats also differ from other hibernators in that they do not restrict hibernation to the winter. In summer, when they rest during the day, their body temperature decreases to the ambient temperature. Thus, they behave as hibernators every day, and the long winter-sleep can be considered as an extension of their daily sleep.

The process of hibernation and the factors that induce or inhibit it vary considerably from one species to another. In nature the beginning of the hibernation cycle is usually associated with the time of the year, but it is not necessarily induced by low temperature or lack of food.

The annual cycle of hibernation is influenced by the duration of the daily light cycle and is also associated with endocrine cycle. For example, it is often impossible to induce even a good hibernator to begin a cycle of hibernation during early summer and especially during the reproductive period.

In arousal the hibernator rapidly activates both heat-conserving and heat-producing mechanisms. The rewarming during arousal is metabolically the most expensive part of the hibernation cycle. However, not only the muscles but also the brown fat is important in the re-warming process. As the brown fat can consume oxygen at a high rate, it is particularly prominent in hibernators and it is considered that mere is a direct connection between hibernation and the amount of brown fat. However, it is most surprising that the brown adipose tissue, which is so important for mammalian hibernators, appears to be absent from a number of birds that regularly hibernate.

Aestivation: Some animals enter the state of torpor that are not necessarily induced by low temperatures and in which most physiological activity appears normal although body temperature may be several degrees below the normal. This is the phenomenon of aestivation. Thus, aestivation (Latin *aestas* = summer) refers to inactivity during the summer and is shown by some mammals and other groups of animals.

It may be applied to snails that become dormant and inactive in response to drought, or it may refer to ground squirrels that during the hottest months disappear into their burrows and remain inactive. However, the Columbia ground squirrel begins to aestivate in the hot month of August and remains inactive throughout the autumn and winter and does not appear again until the following May. Does it aestivate or hibernate, and when does one state change into the other is not fully known.

According to *Schmidt-Nielsen* (1975), no clear physiological distinction can be made between the two states and torpor and hibernation may be treated as one coherent physiological phenomenon.

Regulation of Body Temperature

Human beings are warm-blooded animals and the body temperature is maintained at an average of 36.8°C (98.4°F). In health, variations are usually limited to between 0.5 and 0.75°C, although it may be found that the temperature in the evening is a little higher than in the morning. This is the optimum temperature for the many complex chemical processes to occur. If the temperature is raised the metabolic rate is increased and if it is lowered the rate of metabolism is reduced.

To ensure this constant temperature a fine balance is maintained between heat produced in the body and heat lost to the environment.

Production of Heat Energy

A variety of basic chemical reactions contribute to body heat production at all times. Ingestion of food increases heat production because of the specific dynamic action of the food, but the major source of heat is the contraction of skeletal muscle (Table).

Heat production can be varied by endocrine mechanism in the absence of food intake or muscular exertion. Epinephrine and norepinephrine produce a rapid but short-lived increase in heat production; thyroid hormones produce a slowly developing but prolonged increase. Furthermore, sympathetic discharge is decreased during fasting and increased by feeding.

Body Heat Production and Heat Loss

Body heat is produced by:	
Basic metabolic processes	
Food intake (specific dynamic action)	
Muscular activity	
Percentage of Heat	
Body heat is lost by:	Lost at 21°C
Radiation and conduction	70
Vaporisation of sweat	27
Respiration	2
Urination and defection	1

A source of considerable heat, particularly in infants, is a special type of fat, brown fat, that is located between and around the scapulas and in other parts of the body. This fat has a high rate of metabolism, and its function has been likened to that of an electric blanket.

Heat Decay

Cutaneous Vasodilation: The amount of heat lost from the skin depends to a great extent on the amount of blood in the vessels in the dermis. As the amount of heat produced in the body increases, the artertoles become dilated and more blood pours into the capillary network in the skin. In addition to increasing the amount of sweat produced the *temperature of the skin is raised*. When this happens there is an increase in the amount of heat lost by:

Radiation
Conduction
Convection

In *radiation* the exposed parts of the body radiate heat away from the body.

In *conduction* the clothes in contact with the skin conduct heat away from the body.

In *convection* the air passing over the exposed parts of the body is heated and rises, cool air replaces it and convection currents are set up. Heat is also lost from the clothes by convection.

If the external environmental temperature is low or if heat production is decreased, the blood vessels, under the influence of the sympathetic nerves, constrict, thus decreasing the blood supply to the skin and so preventing heat loss.

Birds have a layer of feathers next to the skin, and most mammals have a significant layer of hair or fur. Heat is conducted from the skin to the air trapped in this layer and from the trapped air to the exterior. When the thickness of the trapped layer is increased by fluffing the feathers or erection of the hairs (horripilation), heat transfer across the layer is reduced and heat losses (or. in a hot environment, heat gains) are decreased.

"Goose pimples" are the result of horripilation in humans; they are the visible manifestation of cold-induced contraction of the piloerector muscles attached to the rather Meager hair supply. Humans usually supplement this layer of hair with a layer of clothes. Heat is conducted from the skin to the layer of air trapped by the clothes, from the inside of the clothes to the outside, and from the outside of the clothes to the exterior.

Dark clothes absorb radiated heat, and light-coloured clothes reflect it back to the exterior.

Sweating: The other major process transferring heat from the body in humans and those animals that sweat is vaporisation of water on the skin and mucous membranes of the mouth and respiratory passages. Vaporisation of 1 g of water removes about 0.6 kcal of heat. A certain amount of water is vaporised at all times. This insensible water loss amounts to 50 mL/h in humans. When sweat secretion is increased, the degree to which the sweat vaporised depends upon the humidity of the environment. It is common knowledge that one feels hotter on a humid day. This is due in part to the decreased vaporisation of sweat, but even under conditions in which vaporisation of sweat is complete, individual in a humid environment feels warmer than an individual in dry environment.

The reason for this difference is not known, but it seems related to the fact that in the humid environment sweat spreads over a greater area of skin before it evaporates. During muscular exertion in a hot environment, sweat secretior reaches values as high as 1600 mL/h, and in a dry atmosphere, most of this sweat is vaporised. Heat loss by vaporisation of water therefore varies from 30 to over 900 kcal.

Panting: Some mammals lose heat by panting. This rapid, shallow breathing greatly increases the amount of water vaporisation in the mouth and respiratory passages and therefore the amount of heat lost. Because the breathing is shallow, it produces relatively little change in the composition of alveolar air.

Temperature-regulating Mechanisms

The reflex and semi-reflex thermoregulatory responses include autonomic, somatic, endocrine, and behavioural changes. One group of responses increases heat loss and decreases heat production ; the other decreases heat loss and increases heat production. In general, exposure to heat stimulates the former group of responses and inhibits the latter, whereas exposure to cold does the opposite.

Curling up "in a ball" is a common reaction to cold in animals and has a counterpart in the position some people assume on climbing into a cold bed. Curling up decreases the body surface exposed to the environment. Shivering is an involuntary response of the skeletal muscles, but cold also causes a semi-conscious general increase in motor activity. Examples include foot stamping and dancing up and down on a cold day. Increased catecholamine secretion is an important endocrine response to cold; adrenal medullectomised rats die faster than normal controls when exposed to cold.

TSH secretion is increased by cold and decreased by heat in laboratory animals, but the change in TSH secretion produced by cold in adult humans is small and of questionable significance. It is common knowledge that activity is decreased in hot weather—the "it's too hot to move" reaction.

Temperature-Regulating Mechanisms

Mechanisms	Effect
Machanisms activated by cold :	
Shivering Hunger Increased voluntary activity Increased secretion of norepinephrine and epinephrine	Increase heat production
Cutaneous vasoconstriction Curling up Horrpilation	Decrease heat loss
Machanisms activated by heat :	
Cutaneous vasodilation Sweating Increased respiration	Increase heat loss
Anorexia Apathy and inertia	Decrease heat production

Thermoregulatory adjustments involve local responses as well as more general responses. When cutaneous blood vessels are cooled they become more sensitive to catercholamines and the arterioles and venules constrict. This local effect of cold directs blood away from the skin.

Another heat-conserving mechanism, that is important in animals living in cold water is heat transfer from arterial to vanous blood in the coming from the extremities (counter-current exchange). This keeps the tips of the extremities cold but conserves body heat.

Nervous Control: The centre controlling temperature is situated in the *cerebrum* and involves a group of nerve cells in the *hypothalamus* called the *heat regulating centre*. There is also a group of nerve cells in the *medulla oblongata* known as the *vasomotor centre* which controls the calibre of the blood vessels, especially the small arteries and the arterioles, and they control and amount of blood which circulates in the capillaries in the dermis. Stimulation of the anterior hypothalamus causes cutaneous vasodilation and sweating. Posterior hypothalamic stimulation causes shivering.

The heat regulating centre and vasomotor centre are thought to be extremely sensitive to the temperature of the blood and any significant *change* stimulates them to activity. From these centres sympathetic nerves convey impulses to the *sweat glands, arterioles* and the *arrector muscles* of the hairs in the skin.

especially the small arteries and the arterioles, and they control and amount of blood which circulates in the capillaries in the dermis. Stimulation of the anterior hypothalamus causes cutaneous vasodilation and sweating. Posterior hypothalamic stimulation causes shivering.

The heat regulating centre and vasomotor centre are thought to be extremely sensitive to the temperature of the blood and any significant change stimulates them to activity. From these centres sympathetic nerves convey impulses to the sweat glands, arterioles and the *arrector muscles* of the hairs in the skin.

8

System of Immunation

The term *immunity* refers to the specific reactions by which the invading microorganisms and their toxic products or other substances foreign to the animal are inhibited, inactivated or destroyed.

In mammals, this system has the remarkable ability to produce *antibodies* against large number of different foreign agents that may invade the body. The antibodies kill the invaders. In addition, the immune system "remembers", and a second exposure to a foreign substance produces a greater and more rapid response.

Plasma Proteins and Immune Reactions: A special group of blood plasma proteins, the *globulins,* plays a major role in the so-called immunological reactions. These include reactions resulting from *blood type* differences and reactions involving the *antigens* and *antibodies.*

An *antigen* is usually a foreign protein molecule (in some cases it is a polysaccharide) which enters the body fluids of an animal as part of an infectious agent and is capable of inducing the production of specific antibodies. An *antibody* is a molecule produced by animals in response to antigen and has the particular property of combining specifically with the antigen which induced its formation. It is a blood protein (globulin) which may make an infected animal immune to a foreign antigen. This response is called the *immune response..* The product of an antigen-antibody reaction, which may also contain components of the complement system, is called *immune complex.*

Cells Involved in Immune Response: All the cells of the immune system are derived from pluripotent stem cells which give rise to two distinct progenitors, one for *myeloid cells* producing phagocytes and other cells, and the other for *lymphoid cells* producing *lymphocytes.* The phagocytes are of two basic kinds, monocytes and polymorphonuclear granulocytes.

The lymphocyte appears to be the key cell in the recognition of antigen and the subsequent immune response. It also appears that there are two populations of lymphocytes, the T-lymphocytes (T for thymus) as T cells differentiate initially in the thymus, and the B- lymphocytes (B for bursa), as in birds B cells differentiate in bursa of Fabricius). But in mammals B cells differentiate in foetal liver, spleen and bone marrow.

The *small lymphocyte* population comprises both B cells and T cells. Although they appear similar, T cells and B cells can be distinguished from one another since they carry different cell surface proteins which act as 'markers'. For example, human T cells, but not B cells, carry a marker which binds to sheep erythrocytes.

B cells or B lymphocytes represent about 5-15 per cent of the circulating lymphoid pool and are classically defined by the presence of endogenously produced immunoglobulins (antibody). During their development, both T and B lymphocytes acquire specific receptors for antigen which commit them to a single antigenic specificity for the rest of their life-span.

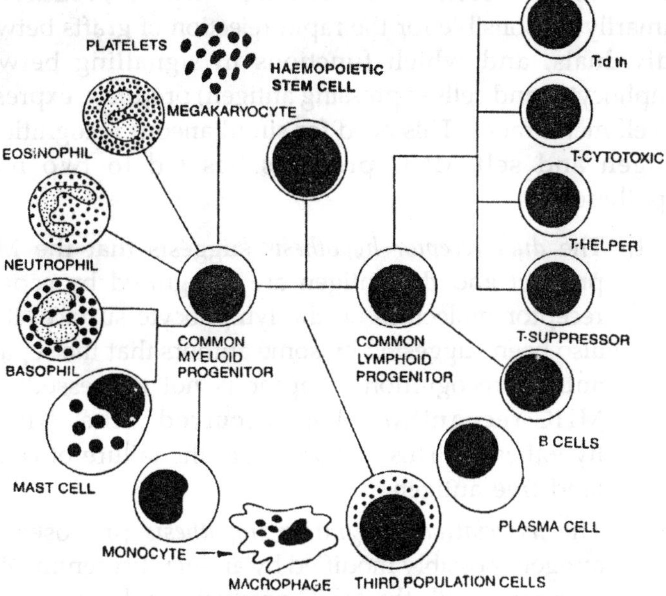

Origin of cells involved in the immune response.

Cell-mediated Protection

The term *'cell-mediated immunity'* (CMI) or *Cellular immunity* was originally coined to describe localised reactions to intracellular pathogens, mediated by T-lymphocytes (and phagocytes) rather than by antibody. However, it is now often used in a more general sense in which antibody plays a subordinate role.

It is well known that T cells have *antigen specificity, i.e.,* they are able to recognise specific antigen. The cell-mediated immune response follows antigen presentation and activation of T cells. Certain T cells (macrophage activating cells) produce *lymphokines.* These molecules activate macrophages to enhance their phagocyte and bactericidal functions.

However, helper (Th) and cytotoxic T cells (Tc) do not bind free antigen. They appear to recognise antigen only in association with MHC (major histocompatibility complex: a genetic region found in all mammals whose products are primarily responsible for the rapid rejection of grafts between individuals, and which functions in signalling between lymphocytes and cells expressing antigen) products, expressed on cell membranes. This need for simultaneous recognition of antigen and self MHC products has led to two major hypotheses:

1. The *dual receptor hypothesis* suggests that the MHC product and the antigen are recognised by separate receptor molecules on the lymphocyte surface. It has also been suggested by some authors that the separate antigen-recognition receptor is not expressed until MHC-recognition has occurred (the 'flasher hypothesis'). This could explain the failure of cells to bind free antigen.
2. The *associative recognition hypothesis* proposes that antigen, possibly modified by antigen-presenting cells, associates with the MHC products, and is recognised by a single T cell receptor. This hypothesis has been modified to give the *altered self hypothesis* which suggests that the self-MHC on the antigen-presenting cell is somehow modified by the presence of antigen. It is proposed that a single T cell receptor recognises both antigen and altered self.

Cytotoxic T cells are activated by antigen and receive help from helper T cells. Helper cells also cooperate with B cells in the production of antibody which may arm cells *(e.g.,* K cells) carrying Fc receptors (Fc is defined as the portion of an antibody which is responsible for binding to antibody receptors on cells and the Clq component of complement). This is called *antibody-dependent cell-mediated cytotoxicity* (ADCC). Thus antibody dependent cellular cytotoxic cells also kill non-specifically, but this occurs via antibodies bound to their target cells. Natural killer (NK) cells act non-specifically, particularly against cellular targets. They may kill tumour cells and virally infected cells and play a role in regulating the immune response.

Finally, it may be mentioned that macrophages play a central role in cell-mediated immunity, because they are involved both in the initiation of responses as antigen-presenting cells, and in the effector phase as inflammatory, tumoricidal and microbicidal cells, in addition to their regulatory functions.

Cellular immunity constitutes a major defence against infections due to viruses, fungi, and a few bacteria such as the tubercle bacillus. It also helps defend against tumours.

Immunoglublins: In 1890, *Von Behring, Buchner,* and others discovered certain proteins when they examined the serum of persons and animals who had been vaccinated against (or who had recovered from) certain infectious diseases. These proteins, now known as *gamma globulins* and called *specific antibodies* or *immunoglobulins,* were found to be defensive only against the specific microorganisms, or their toxins, that evoked those particular antibodies.

For example, diptheria antibodies (antitoxin) "neutralise" diptheria toxin, but not the toxin for tetanus or any other kind. Specific antibodies in the blood were therefore found as the real basis of immunity to disease and gave rise to the doctrine of *humoral* (fluid) *immunity.*

It was also demonstrated by *Tiselius,* the Swedish Biochemist, that the antibody activity of blood was restricted to the gamma globulin fraction of the serum proteins. But in 1954 it was shown that the proteins related to the antibody in structure and function may also be found in other globulin fractions like the beta and even alpha. It has also been shown that all the gamma-globulins need not have the antibody action. Hence, the *term" immunoglobulins"* has been given to these proteins.

There are five distinct classes of immunoglobulins in most higher mammals, including human species. They are IgG, IgA, IgM, IgD and IgE. These differ from each other in size, charge, amino acid composition and carbohydrate content. All immunoglobulins are glycoproteins but the carbohydrate content ranges from 2—3 per cent for IgG to 12—14 per cent for IgM, IgD, and IgE. Within each class two immunologic types have been defined. They are K and L named after Doctor *Korngold* and *Lipari* who first distinguished them. About two thirds of the immunoglobulins of the normal individuals belong to type K or type I, and the rest to type L or Type II.

Structure of Immunoglobulin Molecule: Of all the immunoglobulins, the IgG fraction often accounts for 70 to *75 per cent* of the antibody pool present in normal adult human serum. As such IgG has been more extensively studied. The IgG molecule may be considered as a typical example of the basic antibody structure.

Each molecule of IgG is composed of 4 polypeptide chains—a pair of identical high molecular weight chains, called *heavy* or *H chains,* and a pair of identical low molecular weight chains called *light* or *L chains.* The H chains have a molecular weight of 50,000 to 55,000 and contain about 450 amino acids. The L chains have a molecular weight of 20,000 to 25,000 and contain about 214 amino acids. In associating with each other, the four chains form a molecule whose quaternary structure resembles the letter "Y". Each arm of the Y contains a complete L chain and part of an H chain, while the leg of the Y contains the remaining parts of the H chains. The four polypeptide chains are held together by disulphide (—S—S—) linkages. There is a disulphide link near the C— terminus of each L chain, linking it to the corresponding H chain. Two additional disulphide bridges link the H chains together.

Each of the four polypeptide chains is divided into separate regions called variable and constant *domains.* There are two domains in the L chains (V_L, C_L) and four in the H chains (V_H, C_H1, C_H2, C_H3). Within each of the domains, folding of the polypeptide chain produces two parallel planes each containing several segments with folded beta structure. In each folded beta structure, neighbouring stretches of the polypeptide have opposite polarity, *i.e.,* they are anti-parallel. The two parallel plans in each domain are held together by disulphide bridges and also by Van der Waals interactions between hydrophobic side chains of amino acids in each sheet.

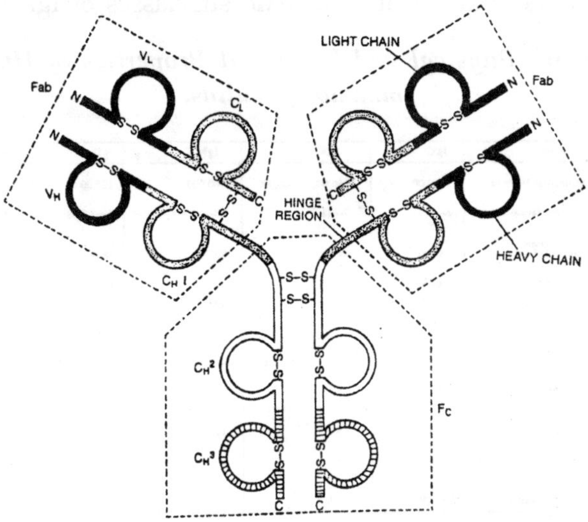

Fundamental architecture of an immunoglobulin molecule.

Enzymatic digestion of the immunoglobulin for subsequent chemical analysis cleaves the molecule in the "hinge" region and produces F_c and F_{ab} fragments. The *variable domain* (portion) of the four polypeptide chains in each immunoglobulin is believed to be responsible for the antibody specificity of the molecule, *i.e.*, the sequence of amino acids in this portion of the molecule determines the spatial configurations for binding a specific antigen.

There are only two types of L chains, either κ (kappa) or λ. (lambda), in IgG and also in other classes of immunoglobulins. They have different but homologous amino acid sequences. However, each class has a unique H chain-γ (gamma) in IgG, α (alpha) in IgA μ (Mu) in IgM, δ (delta) in IgD, ε (epsilon) and in IgE. Like κ and λ chains, the amino acid sequences of the H chains differ but are also homologous. Moreover, there are several subclasses of heavy chains and four are known for

γ-chains, designated $\gamma 1$, $\gamma 2$, $\gamma 3$, and $\gamma 4$ to give IgG_1, IgG_2, etc. The sequences of the H chains within a subclass are also distinct but show very extensive homology in sequence. Similarly there are α_1 and α_2 chain subclasses of IgA.

Important Physical and Biological Properties of Human Immuno globulins.

S. No.	Property	IgG	IgA	IgM	IgD	IgE
1.	Molecular Weight	150,000	160,000 to 500,000	900,000	185,000	200,000
2.	Sedimentation value	6—7 S	7—11 S	18—20 S	6—7 S	8 S
3.	Heavy chains molecular weight	τ 53,000	α 64,000	μ 70,000	δ 58,000	ε 75,000
4.	Light chains molecular weight	κ or λ 22,000	κ or λ 22,000	κ or λ 22,000	κ or λ —	κ or λ 22,000
5.	Molecular formula	$\lambda\kappa_2 \gamma_2$ or $\lambda_2 \gamma_2$	$(\kappa_2 \alpha_2)n$ or $(\lambda_2 \alpha_2)n$ where n = 2, 4	$(\kappa_2 \mu_2)n$ or $(\lambda_2 \mu_2)n$ where n = 5 or multiple of 5	$\kappa_2\delta_2$ or $\lambda_2 \delta_2$	$\kappa_2\varepsilon_2$ or $\lambda_2 \varepsilon_2$
6.	Percent total immunoglobulins	80	13	6	1	0.002
7	Carbohydrate %	2.9	7.5	11.8	—	10.7
8.	Serum concentration (mg/ml)	8—16	1—4	1	0.2	0.03
9.	Half-life (days)	25—35	6—8	9—11	2—3	1—2

κ = kappa

It should be noted that in humans the IgG (but not the IgA or IgM) passes to the foetus and gives passive protection against many infectious agents for upto about 6 months of age.

Other Immunoglobulins and their Applications

Besides IgG, the other major classes of immunoglobulins are IgA and IeM. The minor classes of immunoglobulins found in human plasma are designated as IgD and IgE. It should be noted that IgG, IgD and IgE are monomers. Most IgA and IgM molecules are oligomers, in which n=2 or 4 for IgA and n=5 or a multiple of 5 for IgM. Thus, IgM is a pentamer, whereas IgA may be monomer, dimer, trimer, or tetramer.

IgA: It represents 15-20 per cent of the immunoglobulins in human serum. But it is the principal immunoglobulin in secretions such as saliva, tears and colostrum. It can also be found in the lumen of the intestine, where it is known as *copro antibody.* In secretions, it is generally adimer with an attached S-globulin of about 50,000 molecular weight (called T-piece, transport piece, or secretory piece). The secretory piece is attached to the IeA as it passes through the glandular cell producing the secretion. IgA is thought to be of special importance in the defence of membranes and mouth and vaginal cavities against infections.

IeM: It accounts for about 10 per cent of the immunoglobulin pool. As a pentamer it has increased binding capacity for antigens and thus is extremely efficient. IgM is usually one of the first types of specific antibody to appear in the blood during an infection or after immunisation with most antigens. Thus, it may have a special significance in combating a bacteremia or viremia.

IeD: It constitutes about 1 per cent of the total immunoglobulins. It is known to be present in large quantities on the membrane of many circulating B lymphocytes. IgD molecules were first discovered in patients suffering from a disease of the lymphoid tissues known as multiple myeloma. Their main biological role is not yet described, but they are apparently present in certain intestinal fluids of normal individuals.

IeE: IeE, also called *reagins,* is a trace serum protein found on the surface membrane of basophils and mast cells. IgE may play a role in active immunity to helminthic parasites but in Western countries is more commonly associated with immediate hypersensitivity diseases such as asthma and hayfever.

Humoral (noncellular) Immunity: Humoral means pertaining to the extracellular fluids, including the serum and lymph. B-lymphocytes are primarily responsible for the production of circulating specific antibody, *i.e., humoral immunity.* B-lymphocytes, upon contact with appropriate antigens, are transformed into plasma cells which are responsible for the production of antibody (immunoglobulin), *i.e.,* humoral immunity. Humoral immunity is a major defence against bacterial infections. Thus, humoral immunity is due to the presence of *immunoglobilins* or humoral antibodies, which are a group of glycoproteins present in the serum and tissue fluids of all mammals. Their production is induced when the host's lymphoid system comes into contact with immunogenic foreign molecules (antigens) and they bind specifically to the antigen which induced their formation.

In order to achieve humoral immunity, each immunoglobulin molecule is bifunctional; one region of the molecule (the V domains) is concerned with binding to antigen while a different region (C domains) mediates binding of the immunoglobulin to host tissue. The binding of *antigen* to the *antibody* takes place by the formation of multiple non-covalent bonds between the antigen and amino acids of the binding site. For successful binding there must be suitable atomic groupings on opposing parts of the antigen and antibody. Further, the shape of the combining site must fit the antigen, so that several non-covalent bonds can form simultaneously. If the antigen and the combining site are complementary in this way, there will be sufficient binding energy to resist thermodynamic disruption of the bond. The strength of a single antigen-antibody board is termed the *antibody affinity* and it is produced by summation of the attractive and repulsive forces. A high affinity antibody implies a good fit and conversely, a low affinity antibody implies a poor fit.

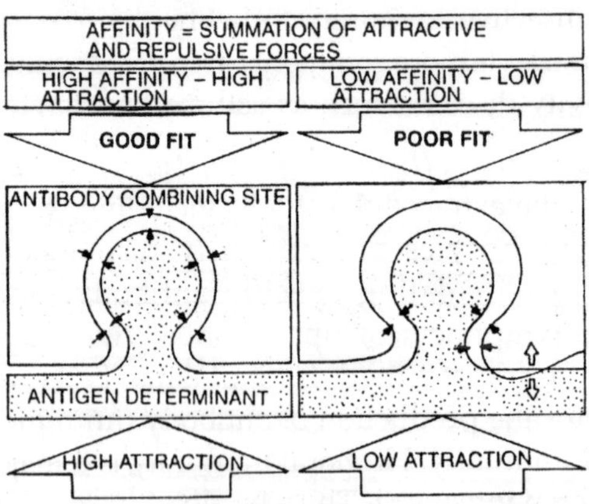

Antibody affinity.

The strength with which a multivalent antibody binds a multivalent antigen is termed *avidity* to differentiate it from the *affinity* of the bond between a single antigenic determinant and an individual combining site. Thus, the *avidity of an antibody* for its antigen is dependent on the affinities of the individual combining sites for the determinants on the antigen, but is greater than the sum of these affinities if both antigen and antibody are multivalent. Antigen-antibody reaction shows a high level of specificity, *i.e.,* the binding sites of antibodies directed against determinants on one antigen are not complimentary to determinants of another antigen. For example, antibodies to a virus like measles will bind to the measles virus and confer humoral immunity to this disease, but will not combine with, or protect against, an unrelated virus such as polio.

The specificity of an antiserum is the result of the summation of actions of the various antibodies in the total population each reacting with a different part of the antigen molecule and even different parts of the same determinant. However, when some determinants of an antigen, A, are shared by another antigen, B, then a proportion of antibodies directed to A will also react with B. This is called *cross-reactivity.* The specificity and cross reactivity expressed by an antiserum are properties which result form the antibody molecules within the serum.

Thus we can say that the antibody recognises the overall configuration of the antigen rather than its chemical composition and it is envisaged that antibodies are directed against particular three-dimensional electron cloud shapes rather than specific chemical structures. In addition, there is frequently an inverse relationship between the charge of an antigen and the antibodies it produces.

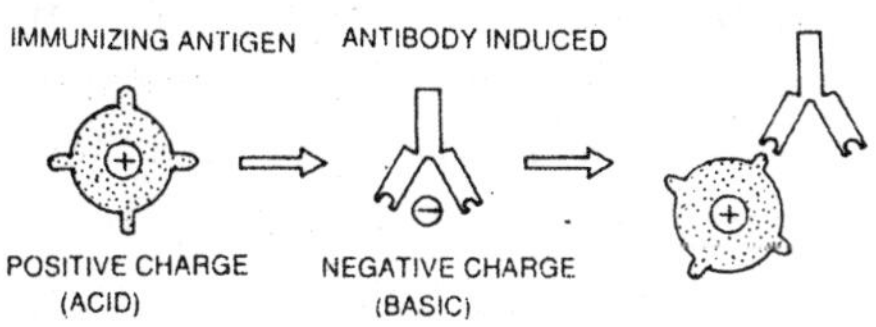

Charge specificity between antigen and antibody.

When an antigen and antibody are allowed to come to equilibrium, it may be written as follows:

$$Ab + Ag \rightleftharpoons AbAg$$

Data from these systems can be analysed by application of the law of mass action to give the affinity constant, K since:

$$K = \frac{[AbAg]}{[Ab][Ag]}$$

where [AbAg] is the concentration of complexed antigen, [Ag] is the concentration of free antigen, and [Ab] is the concentration of free binding sites at equilibrium. When half the binding sites are occupied by antigen, [Ab] = [AbAg] and it follows that, K = 1/[Ag]. In other words, a high affinity antibody only requires a low antigen concentration to achieve binding of antigen to half its combining sites.

Theories of Antibody Formation: Several theories of antibody formation have been proposed. One current theory, called the *clonal selection theory,* is described. According to this theory, recognition of a substance as self or non-self depends on the following mechanism. During foetal development, the lymphoid cells in the primary lymphoid tissues differentiate so that many different cells are obtained, each having one, or perhaps more, recognition sites on its surface. After a short time, thousands of cells, each with one or more different recognition sites, are present in the cell population. As the cells migrate through the developing foetus, some of the cells encounter foetal tissue or soluble antigens which are complementary to their own recognition sites. The two combine, and the lymphoid cell dies. In this way, any lymphoid cells carrying recognition sites for "cell" are destroyed during foetal life. Cells which survive this purge can recognise and react only to non-self. They are free to locate in some lymphoid tissue and produce a *clone, i.e.,* progeny like themselves. It has been postulated that by the time the individual becomes competent immunologically, large numbers of lymphoid cells having between 10,000 and 100,000 different kinds of recognition sites are present in the tissues. It is thought that this is sufficient to recognise almost any non-self antigen to which the host may be exposed.

Thus, when an antigen enters the body and makes contact with a B-or T-lymphocyte bearing the proper recognition site, the lymphocyte is stimulated to reproduce and form a clone (clone means a family of cells or organisms having a genetically identical constitution). Its progeny (B or T) become either *plasma cells* (that secrete humoral antibodies with the same binding specificity as the original clone that differentiated in response to the specific antigen) or *lymphoblasts* (which activate cellular immunity, *i.e.*, phagocytic cells). Thus, B cells become plasma cells and T cells give rise to several types of *effector cells.* They may become (1) *killer T cells,* which eliminate foreign cells by lysis ; (2) *helper T cells,* which aid B cells to differentiate and other T cells to proliferate ; (3) *suppressor T cells,* which suppress the immune response; and (4) *amplifier T cells,* which amplify killer-T-cell response and proliferation. The more the antigenic stimulation of B and T cells, the greater the proliferation of the clones of cells possessing the homologous recognition sites.

Memory of Immunology

The clonal selection theory also accounts for *immunologic memory.* Thus, antibodies are synthesised within a few days after initial exposure to an antigen. After this time, however, reexposure to the antigen elicits a secondary, or anamnestic, response that rapidly results in production of a greater number of antibodies than those produced in the primary response. The *immunologic memory* is based on the fact that the original antigen exposure induces not only effector cells that generate humoral antibodies, but also *memory cells,* which are capable of producing further effector and memory cells after a second exposure to antigen. Unlike effector cells which have lifetimes of days, memory cells survive within the individual for its lifetime and respond rapidly to generate effector cells on any subsequent exposure to antigens.

Mechanism of Immunisation: What does "making immune" mean ? It means the combining of an antibody with an invading antigen. Such a union abolishes the free mobility of the antigen and so prevents the antigen from acting in a damaging manner. The combining process is a lock-and-key reaction similar to the enzyme reaction. The molecular configuration of the antibody protein apparently is such that it "fits into" the configuration of the antigen. In other words, antigen-antibody reactions are specific and any given antigen requires a particular kind of antibody if immunity is to be developed. The specificity has been referred to as an attribute of antigens and antibodies that is a result of mutually corresponding molecular structures called determinants or recognition sites.

The structural resemblance between molecules of an antigen and molecules of the corresponding antibody may be visualised as analogous to the relationship between a mould and its replica or casting. Each is specific for the other. If the antigen molecule is small and contains only one combining site (univalent), then only one antibody molecule can combine with each such antigen molecule (though 2 antigen molecules may combine with each antibody molecule since antibody molecules are typically bivalent). Such antigen-antibody combinations are small and soluble.

The more complex the form of the antigen molecule, the more recognition sites there are for specific combination with antibody, and the larger, more stable and insoluble the combination is. For example, most protein antigens have up to about 5 different determinant groups, each of which evokes an antibody molecule specific for that site. Thus, even though such a multivalent antigen is a pure substance, the antibody produced in response to it is really a mixture of antibody molecules, each specific for a different determinant group on the antigen molecule. Saturation of all the determinant group sites of antigen with specific antibodies forms a large, stable, insoluble and precipitable colloidal complex.

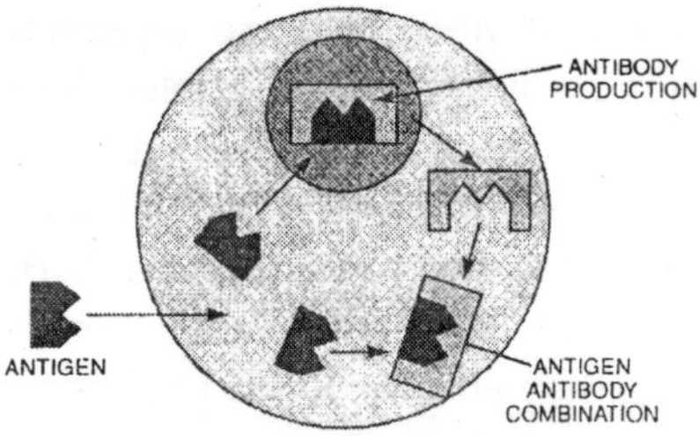

Antigen-antibody reaction.

The following stages are recognised for antigen antibody reactions. The first stage of antigen-antibody reaction is *adsorption* or complex physico-chemical combination (sometimes reversible) between antibodies and antigens. This combination may proceed quite rapidly at 37°C. The bonds between antigen and antibody are non-covalent; they are hydrogen bonds, ionic bonds, etc., and hence are not strong. The second stage of reaction is usually a *visible reaction* (precipitation, cell lysis, or other effect). This stage often develops slowly and may be demonstrated best after 12 to 18 hrs. at 4 to 6°C. However, the presence of electrolytes (NaCl, $MgCl_2$) and certain enzyme-like components of serum *(complement)* is necessary for certain types of second-stage reactions. Complement consists of a complex series of proteins many of which are proteinases. The complement system performs three vital functions: cell activation, cytolysis and opsonisation.

However, animals do not normally possess pre-existing antibodies for all types of antigens which might possibly invade the blood. This means that a specific type of antibody must be manufactured after a given antigen has invaded. Among the organs which produce antibodies appear to be the lymph nodes, the liver, the spleen, and the bone marrow. Among the best understood types of antibodies are: antitoxins and precipitins, agglutinins, cytolysins, immobilising antibodies, and protective and neutralising antibodies.

The combination of antibodies to antigens is manifested in several different ways. One of the most common is the formation of a visible aggregate or precipitate. It is usually called *agglutination* when antigens are aggregated. It is called *precipitation* when soluble antigen molecules become insoluble. The same antibody might be involved in either precipitation or agglutination, depending on whether the determinant groups were attached to a soluble colloid or a particulate substance, such as a whole cell. Frequently when bacteria have combined with antibodies, they are effectively phagocytised by leucocytes and macrophages, a phenomenon called *opsonisation.* Another method of antibody-antigen reaction is the disruption, or *lysis* of cells. This is often observed with erythrocytes (resulting in a release of haemoglobin) and with certain types of bacteria and involves a rupturing of the cells. In addition to the suitable antibodies, the occurrence of opsonisation and lysis requires a component of normal serum called the complement.

Active and Passive Immunisation: The immunity that is conferred by a specific infection or by introducing an antigen through vaccination into the tissues is called *Active immunisation.* Under these conditions the tissues of the host synthesise a relatively large amount of antibodies. *Active natural immunity* means a period of specific immunity that is frequently observed following a natural infection. By this we mean that either the individual is not susceptible to a repeat infection with the same microorganism, or if a repeat infection does occur, the illness is terminated in a much shorter time than initially and usually with less severe symptoms. This type of immunity is long-term or even life long. Vaccinations establish active artificial immunity for periods up to several years or throughout life.

Active artificial immunity depends almost solely upon the induction of specific antibodies, the major exception being when certain live attenuated vaccines are used. In the latter case, both cellular and humoral types of immunity may be elicited.

The principle of *Passive immunisation* involves the use of various animals (rabbit, sheep, horse, guinea pig, etc.) that may be infected and allowed to manufacture specific antibodies against some infectious agent.

Blood serum of these animals, containing the antibodies, may then be used to protect man against the same infectious agent. Diphtheria antitoxin is an example of an immunoglobulin which confers passive immunity. Immunity here is not permanent. Thus, passive immunisation involves antibodies received from another individual. There may be natural or artificial transfer of such antibodies. An example of *passive natural immunity* is the transfer of maternal antibodies via placenta or colostrum. They serve to protect the child for some months after birth. By 3 to 6 months after birth maternal antibodies have largely disappeared and the child becomes susceptible to many infectious diseases. It is therefore, advisable to begin active immunisation of the child early by injections of combined antigens of diphtheria, tetanus, and pertussis, followed by the oral administration of the trivalent oral polio vaccine at 6 months.

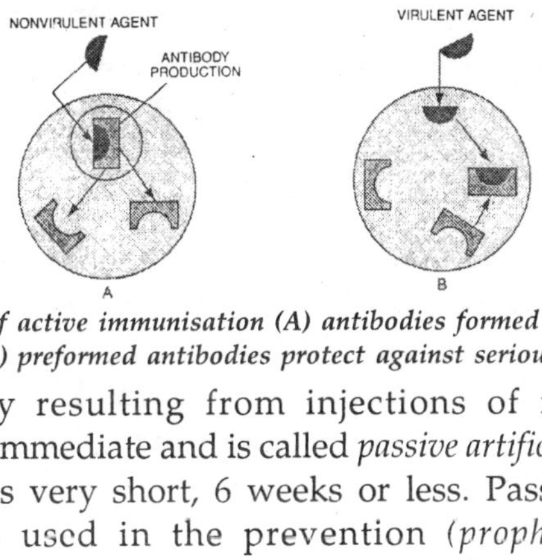

The principle of active immunisation (A) antibodies formed before serious infection, (B) preformed antibodies protect against serious infection.

Immunity resulting from injections of ready-made antibodies is immediate and is called *passive artificial immunity.* Its duration is very short, 6 weeks or less. Passive artificial immunity is used in the prevention (*prophylaxis* from Gr. prophylassein = to guard against) as well as the treatment of diseases. For example, if it is known or suspected by a physician that a person has very recently been exposed, or is likely to become exposed, to certain diseases, a small quantity of serum or some derivative of serum (*e.g.*, purified gamma globulin) containing the appropriate antibodies as a prophylactic measure, is injected.

But to avoid allergic reactions, gamma globulins of human

origin are commonly used for preventive purposes notably against rubella (german measles) and infectious hepatitis. Diseases against which equine serum is occasionally used for prophylaxis are diphtheria and tetanus. Effective human sera are also available for prophylaxis against rabies and pertussis.

Excessive Sensitivity

When an adaptive immune response occurs in an exaggerated or inappropriate form, causing tissue damage, the term *hypersensitivity* is applied. It is a characteristic of the individual and is manifested on second contact with a particular antigen.

Coombs and *Gell* (1963) have described four types of hypersensitivity reaction (type I, II, III and IV), but in practice these types may not necessarily occur in isolation from each other. It should be noted that these reactions are no more than expressions of the beneficial immune responses, acting inappropriately, and sometimes causing inflammatory reactions and tissue damage. The first three types are antibody mediated, and the fourth mediated primarily by T cells and macrophages.

Hypersensitivity—Type I: Type I, or immediate hypersensitivity occurs when an IgE response is directed against innocuous antigens, such as pollen, and the resulting release of pharmacological mediators (such as histamine) by IgE-sensitised mast cells produces an acute inflammatory reaction with symptoms such as asthma or rhinitis. Thus, type I hypersensitivity is characterised by *allergic reacitions* immediately following contact with antigen (allergen).

The term "allergy" was originally coined in 1906 by Von Pirquet meaning 'changed reactivity' of the host when meeting an 'agent' on a second or subsequent occasion. In recent years 'allergy' has become synonymous with Type I hypersensitivity. The immediate anaphylactic response to bee venom is a clearcut example of Type I hypersensitivity, due to the release of pharmacological mediators, including histamine, from mast cells. The reaction can produce generalised anaphylaxis and even death since the allergen is injected into the patient rather than being inhaled.

Hypersensitivity—Type II: Type II, or antibody-dependent cytotoxic hypersensitivity, occurs when antibody (IgG) binds to antigen on cells leading to phagocytosis, killer cell activity, or complement-mediated lysis.

In this form of hypersensitivity, antibody, directed against cell surface or tissue antigens, interacts with molecules of the complement pathways and a variety of effector cells (K cells, platelets, neutrophils, eosinophils, macrophage/monocyte) to bring about damage to those cells and surrounding tissues.

The mechanisms by which phagocytes damage cells in type II hypersensitivity reactions reflect their normal physiological functions in dealing with infectious pathogens.

Hypersensitivity—Type III: Type III, or immune complex mediated hypersensitivity develops when complexes are formed inlarge quantities, or cannot be cleared adequately by the reticuloendothetial system, leading to serum sickness type reactions.

Eeural diseases result from immune complex formation and may affect kidney, skin, lungs, etc. For example, in viral hepatitis, the combined effects of a low grade persistent infection together with a weak antibody response leads to chronic immune complex formation with the eventual deposition of complexes in the tissues.

Hypersensitivity—Type IV: Type IV, or delayed type hypersensitivity (DTH) is manifested when antigen, for example, tubercle bacilli, trapped in a macrophage, cannot be cleared. T lymphocytes are then stimulated to elaborate lymphokines which induce inflammatory reactions and activate and attract macrophages which release mediators. Other aspects of DTH reaction are seen in graft rejection and allergic contact dermatitis.

There are a considerable number of chronic diseases in man which manifest delayed hypersensitivity, and most are due to infectious agents such as mycobacteria, protozoa and fungi. Important diseases in this respect include tuberculosis, leprosy, leishmaniasis, listeriosis, deep fungal infections (*e.g.*, blastomycosis) and helminthic infections (*e.g.*, schistosomiasis). These diseases are caused by pathogens which present a persistant chronic antigenic stimulus. The threat they pose is met by lymphocytes and macrophages. Although these diseases are liable to induce protective immunity, the protective immunity and delayed hypersensitivity do not always coincide.

AIDS and its Destructive Functions

AIDS or Acquired Immune Deficiency Syndrome is a killer disease for which, at present, there is no effective medicine or vaccine. December 1 is celebrated as World AIDS Day to create public awareness regarding this disease and to stress the immediate need to improve monitoring and control the disease. Princess Diana said in an interview in 1993, "The vicious circle of fear, prejudice and ignorance has increased the spread of AIDS to an alarming level. Due to fear and prejudice, many still do not want to listen. After all, AIDS is a killer".

The first cases of AIDS were described in 1981 by *Michael Gottlied* on homosexual men living in Los Angeles, USA. By the end of 1981 epidemiological data indicated that AIDS was an infectious disease transmitted through sexual relations and blood contact. In May 1983, first description of the AIDS infectious agent, the virus LAV (Lymphadenopathic Associated Virus) was given on the basis of the work done by *Prof. Montagnier* and his team at the Pasteur Institute (Paris).

In May 1984, *Prof. Robert Gallo* (USA) and his team isolated the HTLV III (Human T cell leukemia Virus type 3) as infectious agent of AIDS. These two denominations of the same virus (LAV and HTLV III) have now been changed into HIV I (Human Immuno deficiency Virus type 1). In 1985, first routine tests were available for detection of HIV antibodies in blood banks and blood products in North America and Europe. In March 1986, first description and isolation of a second AIDS virus, HIV II a found in West African countries by *Prof. Montagnier* and his team and a Portuguese research team in Lisbon, was made. Due to the high rate of change or maturation that HIV can undergo, new subtypes of this virus are still being discovered. The latest is subtype O in West and Central Africa, taking the total number of HIV subtypes to nearly seven.

It is well known that HIV (human immuno deficiency virus), the retrovirus that causes AIDS, attacks the CD_4 T-cells of the body. These cells are the watch dogs of the human body, the cells that fight an infection. By killing these cells, the body's defence mechanisms are lowered and multiple infections lead to AIDS. While CD_4 is the doorknob to which the HIV virus attaches itself, it has now been discovered that it is the CCR-5 that actually opens the doorknob for the entry of HIV in the body. So, while people with HIV and a normal CCR-5 get full blown AIDS, people with a defective CCR-5 move very slowly towards AIDS. However, people who do not have CCR-5 do not get AIDS at all. Almost in all cases HIV spreads through intravenous drug use, gay sex and sexual activity involving multiple sex partners. In many cases mother-to-child transmission of HIV through breast feeding also occurs.

The HIV viral load tests are conducted by the polymerase chain reaction (PCR) thermocyclen which measures the amount of virus present per ml of blood. In full blown AIDS, the virus load varies between 5 lakh to 32 lakh viral particles per ml.

The first clinical results showing a certain amount of efficiency of AZT *(Azydothymidine)* in the treatment of AIDS patients came at the end of 1986. By 1988 AZT was considered a great hope for HIV-positive people, but its benefits did not last long due to drug resistant HIV strains.

The most recent research from the laboratory of Dr. David

Ho at the Aaron Diamond AIDS Research Centre, New York shows that *"three-drug combination therapy"* could, over three years, flush out the HIV-1 virus from all its hiding places in the body. Referred to as anti-retroviral drugs, the therapy consists of the use of AZT, 3TC and three protease inhibitors. While AZT and 3TC work at the level of reducing the multiplicity of the virus, the protease inhibitor, sequinovir, prevents the development of new virus particles. These expensive drugs suppress HIV to the point that the virus is undetectable. Unfortunately, these therapies can cost upto 25,000 US dollars a year and require patients to follow a strict regimen with nearly 20 pills per day for the rest of their lives. These drugs also have some side effects including nausea, headache, anaemia and toxicity to liver and kidneys. However, no vaccine has been developed against HIV as the virus constantly changes itself.

The latest global statistics show an estimated 30.6 million people with HIV/AIDS infection, of whom 5.8 million have been infected in 1997. AIDS scenario in some of the developing countries is very alarming and children also have been hit by the disease. India has now more than 8 million HIV positive cases. The National AIDS Control Organisation (India) now considers infected children a group large enough to be defined as a separate category. It estimates that 4 per cent per cent of the total AIDS cases in the country are children. AIDS has claimed its first victim in India ın Bombay on June 9,1986. The Indian Council of Medical Research has purchased 2,000 Elisa kits for screening of AIDS patients. One kit could screen 88 patients and costs over 400 U.S. dollars.

Blood Groups and their Characteristics

At least 15 different blood group systems have been recognised in human beings, each system consisting of a gene locus specifying antigens appearing on the erythrocyte surface. The five major blood group systems are: ABO, Rhesus, Kell, Duffy and MN.

The ABO system was discovered by *Landsteiner* (1900). According to him, the human blood was not the same in all individuals and could be classified into four well defined groups (A, B, AB and O). Their percentage or phenotype frequencies are about: A 42 per cent, B 8 per cent, AB 3 per cent and O 47 per cent. Blood grouping is based on the presence or absence of specific *antigens* (agglutinogens) located on the surface of red blood cells.

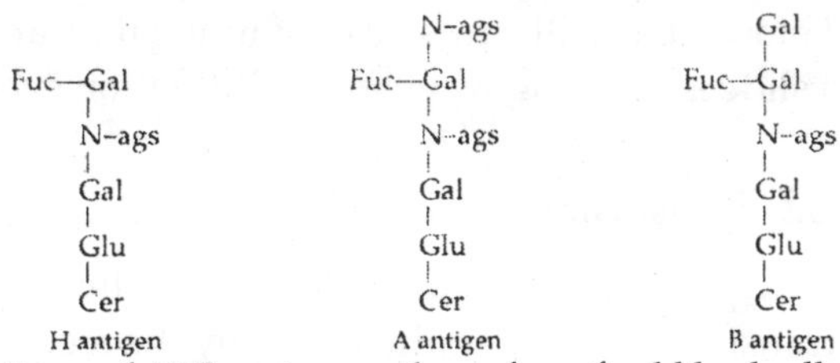

Antigens of ABO system on the surface of red blood cells. Cer, ceramide; Glu, glucose; Gal, galactose; N-ags, N-acetylgalactosamine; Fuc, fucose.

The best known and most important of antigens are the A and B antigens, which are inherited as mendelian dominants. Individuals having blood type A have the A antigen, type B have the B, type AB have both, and type O have neither. The A and B antigens are actually complex oligosaccharides that differ in their terminal sugar. On red cells they are mostly glycosphingolipids, whereas in other tissues *(e.g.,* salivary glands, pancreas, kidney, liver, lungs, testes, etc.) they are glycoproteins. However, H antigen is usually present in individuals of all blood types. An H gene codes for a fucose transferase that puts a *iucose* on the end of H antigen. Individuals who are type A have a gene which codes for a transfrase that catalyses placement of a terminal sugar *N-acetylgalactosamine* on the H antigen. Individuals who are type B have a gene which codes for a transferase that places *galactose* as terminal sugar. Individuals who are type AB have both transferases, *i.e.* having both N-acetylgalactosamine and galactose as terminal sugar. Individuals who are type O have neither, so the H antigen persists.

Antibodies against red cell antigens are called *agglutinins* and are present in plasma. Thus, type O contains antibodies *a* and *b*; type A contains antibody *b* (Anti-B); type B contains antibody *a* (Anti-A) and type AE contains no antibodies. In these groups normally no blood agglutinates itself because the corresponding antigen and antibody, as, for example, A and *a*, or B and *b* are not present together, for *a* is anti-A and *b* is anti-B. On the other hand A is compatible with *b* and B is compatible with *a*. Blood type is decided by mixing an individual's red blood cells with antisera containing the various agglutinins on a slide and seeing whether agglutination occurs.

Theory of Blood Transfusion

The basic principle of transfusion states that if the cells of the blood to be transfused will be agglutinated by the recipients plasma, the transfusion must not be made. Thus, since the cells of type A blood are agglutinated by B plasma, A blood cannot be given to a B patient. On the other hand, the cells of O blood are not agglutinated by any plasma, therefore O blood can be given to patients with other types of blood. Accordingly, the individuals having O blood are termed *universal donors.* The plasma of AB blood does not agglutinate the cells of any blood. Hence, the individuals having this type of blood are termed *universal recipients* (Table). However, in actual practice doctors insist on giving the blood of the same type as the patient's blood to prevent any possibility of incompatibility.

Main Blood Groups with their Antigens and Antibodies and their Possible Combinations in Blood Transfusions.

S.No.	*Blood*	*Antigens on erythrocyte surface*	*Antibodies in serum*	*Can give blood to*	*Can receive blood from*
1.	AorA$_b$	A	*b* (Anti-B)	A,AB	O,A
2.	B or B$_a$	B	*a* (Anti-A	B,AB	O, B
3.	AB	A and B	none	AB	all
4.	O or Oab	none	*a, b* (Anti-A & Anti-B)	all	O

Transfusion Activities

Transfusion of blood into a recipient who has antibodies to those red cells produces an immediate transfusion reaction. For example, when the plasma of a type A individual is mixed with type B red cells, the anti-B antibodies cause the type B red cells to clump (agglutinate). Clinical indications of transfusion reactions are fever, hypotension, lower back pain, feeling of chest compression, and nausea and vomiting.

The severity of the reaction depends on the class and amounts of the antibodies involved. Antibodies to the ABO system antigens are usually of the IgM class; they cause agglutination, complement activation and intravascular haemolysis.

Other blood groups induce IgG antibodies, and although these agglutinate the cells less well than IgM antibodies, they activate Type II hypersensitivity mechanisms and cause red cell destruction. The cell destruction may cause circulatory shock, and the released contents of erythrocytes can produce acute tubular necrosis of the kidneys.

Transfusion reactions to incompatible blood may also develop over days or weeks as antibodies to the foreign cells are produced, resulting in anaemia and jaundice.

Rh Factor

The Rh antigen was first discovered by *Landsteiner* and *Weiner* (1940) in the rhesus monkey. Later on it was shown that most human beings also have this factor and are said to be *Rh-positive* (RhD^+), while only about 15 per cent are *Kh-negative* (RhD^-) Rh-positive and Rh-negative bloods are incompatible and cannot be mixed.

Since the discovery of Rh factor, it has been established that there are no less than 6 different antigens (C or c, D or *d,* E or *e)* involved in the Rh system. The most important of which is referred to as the *antigen* D or the *Rh factor,* which is usually of serious clinical significance as it evokes a stronger reaction in an incompatible recipient than the other Rhesus antigens. There is no antibody that normally accompanies the *Rh* antigen as is true of the antigens of ABO blood groups. If the blood of an Rh-positive person is transfused into an Rh-negative person, the Rh antigen will, however, cause the formation of antibodies. If at some subsequent time another transfusion of Rh-positive blood is made to this Rh-negativeperson, then the Rh antigen reacts with the antibodies, causing a serious transfusion reaction, agglutination, which may result to death. Another serious result of Rh-incompatibility, where RhD^+ is the most commonly involved antigen, may be caused when an Rh-negative pregnant woman carries an Rh-positive child. Rh-positive erythrocytes may pass from the bloodstream of the foetus into the mother's blood where they stimulate production of IgG antibodies.

These antibodies of the mother may diffuse into the bloodstream of the foetus and cause destruction of foetal erythrocytes, a characteristic form of anaemia known as *erythroblastosis foetalis* or *haemolytic disease of the newborn* (HDNB). This results in varying degrees of injury to the foetus, depending upon the concentration of the antibodies present. The effects range all the way from a very mild anaemia to more severe anaemic conditions, various structural abnormalities, abnormalities of the nervous system, sometimes causing mental deficiency, to miscarriage or still birth.

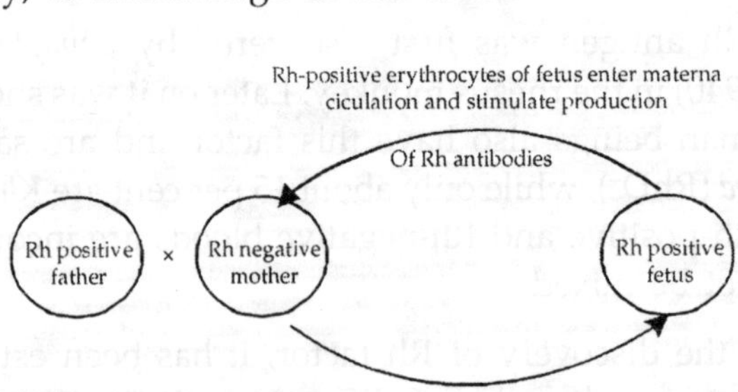

Antibodies produced by mother enter fetal circulation and cause erythroblastosis

However, for reasons as yet unexplained, the transfer of erythrocytes and antibodies across the placenta occurs only in a small percentage of cases. Thus, in only one mating out of 28 between Rh-positive males and Rh-negative females is there any sign of the effects of Rh incompatibility. The first and the second child born of parents with Rh incompatible blood groups usually escape harm. But by the time the third child is conceived there may be sufficient antibodies in the mother's blood to produce serious effects like the death of the newborn infant or abortion of the foetus.

M and N Factors: The two antigens, M and N, give rise to three blood groups called M, N and MN. Their percentage is 28 per cent, 50 per cent and 22 per cent respectively. The presence of these antigens is determined by two genes that reside in the nuclear chromosomes.

Use of M and N Factors for Establishing Parentage.

If child is	*The mother*	*The father*	
		Must be	*cannot be*
M	must be M or MN	M or MN	N
N	must be N or MN	N or MN	M
MN	may be M	N or MN	M
	N	M or MN	N
	MN	M, N, or MN	

Differing from the ABO blood groups, the antigens M and N associated with the red blood corpuscles are not accompanied by antibodies in the plasma. Hence, in blood transfusions these antigens are not taken into consideration. However, they are of great value in deciding the parentage of a child when this is in dispute.

9

Survival and Growth (Plants)

Plant survival and growth is critically influenced by abiotic factors including water, wind, and light. Hormones play an important role in the internal signalling that brings about environmental responses, and is keyed in many ways to the environment. The effect of the local environment on plant growth also accounts for much of the variation in adult form within a species. Precisely regulated responses to the environment not only allow a plant to survive from day to day but also determine when a flowering plant will produce a flower. All organisms sense and interact with their environment. This is particularly true of plants.

Tropisms

Growth patterns in plants are often guided by environmental signals. Tropisms (from trope, the Greek word for "turn") are positive or negative growth responses of plants to external stimuli that usually come from one direction. Some responses occur independently of the direction of the stimuli and are referred to as nastic movements. For example, a tendril of a pea plant will always coil in one direction when touched.

Tropisms, on the other hand, are directional and offer significant compensation for the plant's inability to get up and walk away from unfavourable environmental conditions.

Tropisms contribute the variety of branching patterns we see within a species. Three major classes of plant tropisms: phototropism, gravitropism, and thigmotropism. Tropisms are particularly intriguing because they challenge us to connect environmental signals with cellular perception of the signal, transaction into biochemical pathways, and ultimately an altered growth response.

Phototropism : Phototropic responses involve the bending of growing stems and other plant parts towards sources of light. In general, stems are positively phototropic, growing towards a light source, while most roots do not respond to light or, in exceptional cases, exhibit only a weak negative phototropic response. The phototropic reactions of stems are clearly of adaptive value, giving plants greater exposure to available light. They are also important in determining the development of plant organs and, therefore, the appearance of the plant.

Individual leaves may display phototropic responses. The position of leaves is important to the Photosynthetic efficiency of the plant. A plant hormone called auxin is probably involved in most, if not all, of the phototropic growth responses of plants. The first step in a phototropic response is perceiving the light. Photo-receptors perceive different wavelengths of light with blue and red being the most common. Blue light receptors are being characterised and we are beginning to understand how plants "see blue."

Much more is known about "seeing red" and translating that perception into a signal transaction pathway leading to an altered growth response. Plants contain a pigment, phytochrome, which exists in two interconvertible forms, Pr and Pfr. In the first form, phytochrome absorbs red light; in the second, it absorbs far-red light. When a molecule of

Pr absorbs a photon of red light (660 nm), it is instantly converted into a molecule of Pfr, and when a molecule of Pfr absorbs a photon of far-red light (730 nm), it is instantly converted to Pr. Pfr is biologically active and Pr is biologically inactive.

In other words, when Pfr is present, a given biological reaction that is affected by phytochrome will occur. When most of the Pfr has been replaced by Pr, the reaction will not occur. Phytochrome as a single molecule, it is important to note that several different phytochromes have now been identified that appear to have specific biological functions.

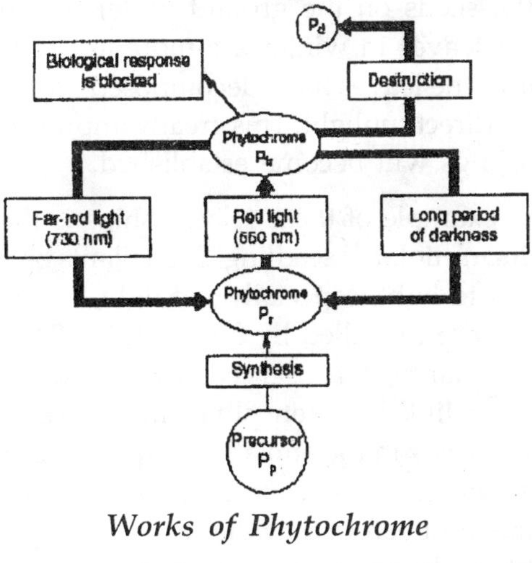

Works of Phytochrome

Phytochrome is a light receptor, but it does not act directly to bring about reactions to light. The existence of phytochrome was conclusively demonstrated in 1959 by Harry A. Borthwick and his collaborators at the U.S. Department of Agriculture Research Centre at Beltsville, Maryland. It has since been shown that the molecule consists of two parts: a smaller one that is sensitive to light and a larger portion that is a protein. The protein component initiates a signal transaction leading to a particular tropism.

The phytochrome pigment is blue, and its light sensitive portion is similar in structure to the phycobilins that occur in

Cyanobacteria and red algae. Phytochrome is present in all groups of plants and in a few genera of green algae, but not in bacteria, fungi, or protists (other than the few green algae). It is likely that phytochrome systems for measuring light evolved among the green algae and were present in the common ancestor of the plants.

Phytochrome is involved in many plant growth responses. For example, seed germination is inhibited by fared light and stimulated by red light in many plants. Because chlorophyll absorbs red light strongly but does not absorb far-red light, light passing through green leaves inhibits seed germination. Consequently, seeds on the ground under deciduous plants that lose their leaves in winter are more apt to germinate in the spring after the leaves have decomposed and the seedlings are exposed to direct sunlight. This greatly improves the chances that the seedlings will become established.

A second example of these relationships is the elongation of the shoot in an etiolated seedling. Such plants become normal when exposed to light, especially red light, but the effects of such exposure are cancelled by far-red light. This indicates a relationship similar to that observed in seed germination. There appears to be a link between phytochrome light perception and brassinosteroids in the etiolation response. Etiolation is an energy conservation strategy to help plants growing in the dark reach the light before they die. They don't green-up until there is light, and they divert energy to growing as tall as possible through internode elongation. The de-etiolated Arabidopsis mutant has a poor etiolation response.

It does not have elongated internodes and greens up a bit in the dark. It turns out that det2 mutants are defective in an enzyme necessary for brassinosteroid biosynthesis. Researchers suspect that brassinosteroids play a role in how plants respond to light through phytochrome. Thus, because det2 mutants lack brassinosteroids, they do not respond to light, or lack of light, as normal plants do, and the det2 mutants grow normally in the dark.

Red and far-red light also are used as signals for plant spacing. The closer plants are together, the more likely they are to grow tall and try to outcompete others for the sunshine. Plants somehow measure the amount of far-red light being bounced back to them from neighbouring trees. If their perception is messed up by putting a collar around the stem with a solution that blocks light absorption, the elongation response is no longer seen.

Gravitropism : When a potted plant is tipped over, the shoot bends and grows upward. The same thing happens when a storm pushes over plants in a field. These are examples of gravitropism, the response of a plant to the gravitational field of the earth. Brown algae orient their first cell division so the rhizoid grows downward. Rhizoids also develop away from a unilateral light source. Separating out phototropic effects is important in the study of gravitropisms.

Gravitropic responses are present at germination when the root grows down and the shoot grows up. Why does a shoot have a negative gravitropic response (growth away from gravity), while a root has a positive gravitropic response? The opportunity to experiment on the space shuttle in a gravity-free environment has accelerated research in this area. Auxins play a primary role in gravitropic responses, but they may not be the only way gravitational information is sent through the plant.

When John Glenn made his second trip into space, he was accompanied by an experiment designed to test the role of gravity and electrical signalling in root bending. Analysis of gravitropic mutants is also adding to gravitropism. There are four steps that lead to a gravitropic response:

- Gravity is perceived by the cell
- Signals form in the cell that perceives gravity
- The signal is transduced intra- and intercellularly
- Differential cell elongation occurs between cells in the "up" and "down" sides of the root or shoot.

One of the first steps in perceiving gravity is that amyloplasts, plastids that contain starch, sink towards the gravitational field. These may interact with the cytoskeleton, but the net effect is that auxin becomes more concentrated on the lower side of the stem axis than on the upper side.

The increased auxin concentration on the lower side in stems causes the cells in that area to grow more than the cells on the upper side. The result is a bending upward of the stem against the force of gravity in other words there is a negative gravitropic response. Such differences in hormone concentration have not been as well documented in roots. Nevertheless, the upper sides of roots oriented horizontally grow more rapidly than the lower sides, causing the root ultimately to grow downward; this phenomenon is known as positive gravitropism. In shoots, the gravity sensing cells are in the endoderm.

Mutants like scarecrow and short root in Arabidopsis that lack normal endodermal development fail to have a normal gravitropic response. These endodermal cells are the sites of the amyloplasts in the stems. In roots, the gravity-sensing cells are located in the root cap and the cells that actually do the asymmetric growth are in the distal elongation zone which is closest to the root cap. How the information gets transferred over this distance is an intriguing problem. Auxin may be involved, but when auxin transport is suppressed, there is still a gravitropic response in the distal elongation zone.

Some type of electrical signalling involving membrane polarisation has been hypothesised and this was tested aboard the space shuttle. So far the verdict is still not in on the exact mechanism. It may surprise you to learn that in tropical rain forests, roots of some plants may grow up the stems of neighbouring plants, instead of exhibiting the normal positive gravitropic responses typical of other roots. The rainwater dissolves nutrients, both while passing through the lush upper canopy of the forest, and also subsequently as it trickles down tree trunks. Such water functions as a more reliable source of nutrients for the roots than the nutrient-poor soils in which the plants are anchored.

Explaining this in terms of current hypotheses is a challenge. It has been proposed that roots are more sensitive to auxin than shoots and that auxin may actually inhibit growth on the lower side of a root, resulting in a positive gravitropic response. Perhaps in these tropical plants, the sensitivity to auxin in roots is reduced.

Thigmotropism : Thigmotropism is a name derived from the Greek root thigma, meaning "touch." A thigmotropism is a response of a plant or plant part to contact with the touch of an object, animal, plant, or even the wind. When a tendril makes contact with an object, specialised epidermal cells, whose action is not clearly understood, perceive the contact and promote uneven growth, causing the tendril to curl around the object, sometimes within as little as 3 to 10 minutes. Both auxin and ethylene appear to be involved in tendril movements, and they can induce coiling in the absence of any contact stimulus.

In other plants, such as clematis, bindweed, and dodder, leaf petioles or unmodified stems twine around other stems or solid objects. Again, Arabidopsis is proving valuable as a model system. A gene has been identified that is expressed in 100-fold higher levels 10 to 30 minutes after touch. Given the value of a molecular genetics approach in dissecting the pathways leading from an environmental signal to a growth response, this gene provides a promising first step in understanding how plants respond to touch.

Other Tropisms : The tropisms just discussed are among the best known, but others have been recognised. They include electrotropism (responses to electricity); chemotropism (response to chemicals); traumotropism; thermotropism (response to temperature); aerotropism (response to oxygen); skototropism (response to dark); and geomagnetotropism (response to magnetic fields). Roots will often follow a diffusion gradient of water coming from a cracked pipe and enter the crack. Some call such growth movement hydrotropism, but there is disagreement whether responses to water and several other "stimuli" are true tropisms.

Dormancy : Sometimes modifying the direction of growth is not enough to protect a plant from harsh conditions. The ability to cease growth and go into a dormant stage provides a survival advantage. The extreme example is seed dormancy, but there are intermediate approaches to waiting out the bad times as well. Environmental signals both initiate and end dormant phases in the life of a plant.

In temperate regions, when freezing temperatures and the accompanying unavailability of water make it impossible for plants to grow. During this season, buds of deciduous trees and shrubs remain dormant, and apical meristems remain well protected inside enfolding scales. Perennial herbs spend the winter underground as stout stems or roots packed with stored food. Many other kinds of plants, including most annuals, pass the winter as seeds.

In some seasonally dry climates, seed dormancy occurs primarily during the dry season, often the summer. Rainfalls trigger germination when conditions for survival are more favourable. Annual plants occur frequently in areas of seasonal drought. Seeds are ideal for allowing annual plants to bypass the dry season, when there is insufficient water for growth. When it rains, they can germinate and the plants can grow rapidly, having adapted to the relatively short periods when water is available.

These include the leaching from the seed coats of chemicals that inhibit germination, or mechanically cracking the seed coats, a procedure that is particularly suitable for promoting growth in seasonally dry areas. Whenever rains occur, they will leach out the chemicals from the seed coats, and the hard coats of other seeds may be cracked when they are being washed down along temporarily flooded arroyos.

Seeds may remain dormant for a surprisingly long time.

Many legumes (plants of the pea and bean family, Fabaceae) have tough seeds that are virtually impermeable to water and oxygen. These seeds often last decades and even longer without special care; they will eventually germinate when their seed coats have been cracked and water is available. Seeds that are thousands of years old have been successfully germinated.

A period of cold is necessary before some kinds of seeds will germinate. The seeds of other plants will germinate only when adequate water is available and the temperatures are relatively high. For this reason, certain weeds germinate and grow in the cooler part of the year and others in the warmer part of the year.

Similarly, a period of cold is needed before the buds of some trees and shrubs will break dormancy and develop normally. For this reason, many plants that normally grow in temperate regions do not thrive in warmer regions near the equator, because even at high elevations in the tropics it still does not get cold enough, and the day-length relationships are different from those of temperate regions.

Growth Guiding Hormones

Plant Hormones : While initial responses of plants to environmental signals may rely primarily on electrical signalling, longer-term responses that alter morphology rely on complex physiological networks. Many internal signalling pathways involve plant hormones. Hormones are involved in responses to the environment, as well as internally regulated development. Hormones are chemical substances produced in small, often minute, quantities in one part of an organism and then transported to another part, where they bring about physiological or developmental responses. The activity of hormones results from their capacity to stimulate certain physiological processes and to inhibit others. How they act in a particular instance is influenced both by the hormone and the tissue that receives the message.

In animals, hormones are usually produced at definite sites, usually organs. In plants, hormones are not produced in specialised tissues but, instead, in tissues that also carry out other, usually more obvious, functions. There are seven major kinds of plant hormones: auxin, cytokinins, gibberellins, brassinosteroids, oligosaccharins, ethylene, and abscisic acid. Current research is focused on the biosynthesis of hormones and on characterising the hormone receptors that trigger signal transaction pathways. Much of the molecular basis of hormone function remains enigmatic.

Auxin : More than a century ago, an organic substance known as auxin became the first plant hormone to be discovered. Auxin increases the plasticity of plant cell walls and is involved in elongation of stems. Cells can enlarge in response to changes in turgor pressure when their cell walls have enhanced plasticity from auxin action. The discovery of auxin and its role in plant growth is an elegant example of thoughtful experimental design. The historical story is recounted here for that reason. Recent efforts have uncovered an auxin receptor. Transport mechanisms are also being unravelled. As with all the classes of hormones, at a cellular and molecular level, how hormones regulate growth and development.

In his later years, the great evolutionist, Charles Darwin, became increasingly devoted to the study of plants. In 1881, he and his son Francis published a book called The Power of Movement of Plants. In this book, the Darwins reported their systematic experiments on the response of growing plants to light responses that came to be known as phototropisms. They used germinating oat (Avena sativa) and canary grass (Phalaris canariensis) seedlings in their experiments and made many observations in this field.

Charles and Francis Darwin knew that if light came primarily from one direction, the seedlings would bend strongly towards it. If they covered the tip of the shoot with a thin glass tube, the shoot would bend as if it were not covered. However, if they used a metal foil cap to exclude light from the plant tip, the shoot would not bend.

They also found that using an opaque collar to exclude light from the stem below the tip did not keep the area above the collar from bending. In explaining these unexpected findings, the Darwins hypothesised that when the shoots were illuminated from one side, they bent towards the light in response to an "influence" that was transmitted downward from its source at the tip of the shoot. For some 30 years, the Darwins' perceptive experiments remained the sole source of information about this interesting phenomenon.

Then Danish plant physiologist Peter Boysen-Jensen and the Hungarian plant physiologist Arpad Paal independently demonstrated that the substance that caused the shoots to bend was a chemical. They showed that if the tip of a germinating grass seedling was cut off and then replaced with a small block of agar separating it from the rest of the seedling the seedling would grow as if there had been no change. Something evidently was passing from the tip of the seedling through the agar into the region where the bending occurred.

On the basis of these observations under conditions of uniform illumination or of darkness, Paal suggested that an unknown substance continually moves down from the tips of grass seedlings and promotes growth on all sides. Such a light pattern would not, of course, cause the shoot to bend.

Then, in 1926, Dutch plant physiologist Frits Went carried Paal's experiments an important step further. Went cut off the tips of oat seedlings that had been illuminated normally and set these tips on agar. He then took oat seedlings that had been grown in the dark and cut off their tips in a similar way. Finally, Went cut tiny blocks from the agar on which the tips of the light-grown seedlings had been placed and placed them off-centre on the tops of the decapitated dark-grown seedlings.

Even though these seedlings had not been exposed to the light themselves, they bent away from the side on which the agar blocks were placed. Went then put blocks of pure agar on the decapitated stem tips and noted either no effect or a slight bending towards the side where the agar blocks were placed. Finally, Went cut sections out of the lower portions of the lightgrown seedlings to see whether the active principle was present in them. He placed these sections on the tips of decapitated, dark-green oat seedlings and again observed no effect.

As a result of his experiments, Went was able to show that the substance that had diffused into the agar from the tips of light-grown oat seedlings could make seedlings curve when they otherwise would have remained straight. He also showed that this chemical messenger caused the cells on the side of the seedling into which it flowed to grow more than those on the opposite side. In other words, it enhanced rather than retarded cell elongation. He named the substance that he had discovered auxin, from the Greek word auxein, which means "to increase."

Went's experiments provided a basis for understanding the responses that the Darwins had obtained some 45 years earlier. The oat seedlings bent towards the light because of differences in the auxin concentrations on the two sides of the shoot. The side of the shoot that was in the shade had more auxin, and its cells therefore elongated more than those on the lighted side, bending the plant towards the light.

Effects of Auxins : Auxin acts to adapt the plant to its environment in a highly advantageous way. It promotes growth and elongation and facilitates the plant's response to its environment. Environmental signals directly influence the distribution of auxin in the plant. How does the environment specifically, light, exert this influence? Theoretically, it might destroy the auxin, decrease the cells' sensitivity to auxin, or cause the auxin molecules to migrate away from the light into the shaded portion of the shoot. This last possibility has proved to be the case.

In a simple but effective experiment, Winslow Briggs inserted a thin sheet of mica vertically between the half of the shoot oriented towards the light and the half of the shoot oriented away from it. He found that light from one side does not cause a shoot with such a barrier to bend. When Briggs examined the illuminated plant, he found equal auxin levels on both the light and dark sides of the barrier. He concluded that a normal plant's response to light from one direction involves auxin migrating from the light side to the dark side, and that the mica barrier prevented a response by blocking the migration of auxin.

The effects of auxin are numerous and varied. Auxin promotes the activity of the vascular cambium and the vascular tissues. Also, auxins are present in pollen in large quantities and play a key role in the development of fruits. Synthetic auxins are used commercially for the same purpose. Fruits will normally not develop if fertilization has not occurred and seeds are not present, but frequently they will if auxins are applied. Pollination may trigger auxin release in some species leading to fruit development occurring even before fertilization.

How Auxin Works

In spite of this long history of research on auxin, its molecular basis of action has been an enigma. The chemical structure of IAA resembles that of the amino acid tryptophan, from which it is probably synthesised by plants. Unlike animal hormones, a specific signal is not sent to specific cells, eliciting a predictable response. There are most likely multiple auxin perception sites. Auxin is also unique among the plant hormones in that it is transported towards the base of the plant.

Two families of genes have been identified in Arabidopsis that are involved in auxin transport. For example, one protein is involved in the top to bottom transport of auxin; while two other proteins function in the root tip to regulate the growth response to gravity. One of the downstream effects of auxin is an increase in plasticity of the plant cell wall.

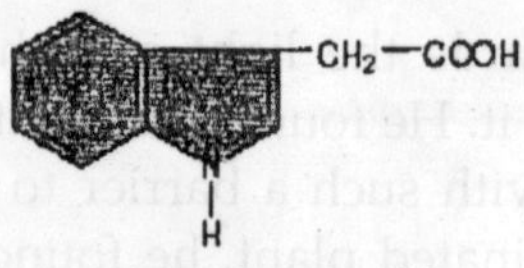

(a) IAA (Indoleacetic acid)

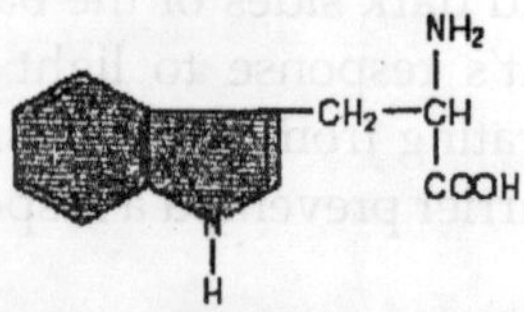

(b) Tryptophan

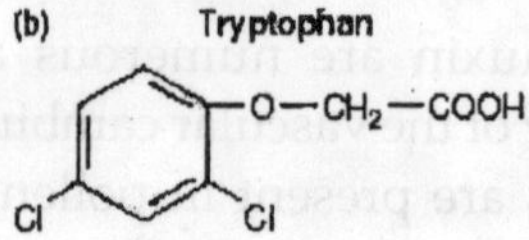

Auxins, (a) Indoleacetic acid (IAA), the principal naturally occurring auxin. (b) Tryptophan, the amino acid from which plants probably synthesise IAA. (c) Dichlorophenoxyacetic acid (2,4-D), a synthetic auxin, is a widely used herbicide.

This will only work on young cell walls without extensive secondary cell wall formation. A more plastic wall will stretch more while its protoplast is swelling during active cell growth. The acid growth hypothesis provides a model linking auxin to cell wall expansion. Auxin causes responsive cells to release hydrogen ions into the cell wall. This decreases the pH which activates enzymes that can break bonds between cell wall fibres. Remember that different enzymes operate optimally at different pHs. This hypothesis has been experimentally supported in several ways. Buffers that prevent cell wall acidification block cell expansion. Other compounds that release hydrogen ions from the cell can also cause cell expansion. The movement of hydrogen ions has been observed in response to auxin treatment. This hypothesis explains the rapid growth response. There are also delayed responses which most likely involve auxin-stimulated gene expression.

Synthetic Auxins : Synthetic auxins such as NAA (naphthalene acetic acid) and IBA (indolebutyric acid) have many uses in agriculture and horticulture. One of their most important uses is based on their prevention of abscission, the process that causes a leaf or other organ to fall from a plant. Synthetic auxins are used to prevent fruit drop in apples before they are ripe and to hold berries on holly that is being prepared for shipping. Synthetic auxins are also used to promote flowering and fruiting in pineapples and to induce the formation of roots in cuttings. Synthetic auxins are routinely used to control weeds. When used as herbicides, they are applied in higher concentrations than IAA would normally occur in plants. One of the most important synthetic auxin herbicides is 2,4-dichlorophenoxyacetic acid, usually known as 2,4-D. It kills weeds in grass lawns by selectively eliminating broad-leaved dicots. The stems of the dicot weeds cease all axial growth.

The herbicide 2,4,5-trichlorophenoxyacetic acid, better known as 2,4,5-T, is closely related to 2,4-D. 2,4,5-T was widely used as a broad-spectrum herbicide to kill weeds and seedlings of woody plants. It became notorious during the Vietnam War as a component of a jungle defoliant known as Agent Orange and was banned in 1979 for most uses in the United States.

When 2,4,5-T is manufactured, it is unavoidably contaminated with minute amounts of dioxin. Dioxin, in doses as low as a few parts per billion, has produced liver and lung diseases, leukaemia, miscarriages, birth defects, and even death in laboratory animals. Vietnam veterans and children of Vietnam veterans exposed to Agent Orange have been among the victims.

Cytokinins : Cytokinins comprise another group of naturally occurring growth hormones in plants. Studies by Gottlieb Haberlandt of Austria around 1913 demonstrated the existence of an unknown chemical in various tissues of vascular plants that, in cut potato tubers, would cause parenchyma cells to become meristematic, and would induce the differentiation of a cork cambium. The role of cytokinins, active components of coconut milk, in promoting the differentiation of organs in masses of plant tissue growing in culture later led to their discovery. Subsequent studies have focused on the role cytokinins play in the differentiation of tissues from callus. A cytokinin is a plant hormone that, in combination with auxin, stimulates cell division and differentiation in plants. Most cytokinins are produced in the root apical meristems and transported throughout the plant.

Developing fruits are also important sites of cytokinin synthesis. In mosses, cytokinins cause the formation of vegetative buds on the gametophyte. In all plants, cytokinins, working with other hormones, seem to regulate growth patterns. All naturally occurring cytokinins are purines that appear to be derivatives of, or have molecule side chains similar to, those of adenine. Other chemically diverse molecules, not known to occur naturally, have effects similar to those of cytokinins. Cytokinins promote growth of lateral buds into branches; though, along with auxin and ethylene, they also play a role in apical dominance (the suppression of lateral bud growth). Conversely, cytokinins inhibit formation of lateral roots, while auxins promote their formation. As a consequence of these relationships, the balance between cytokinins and auxin, along with other factors, determines the appearance of a mature plant.

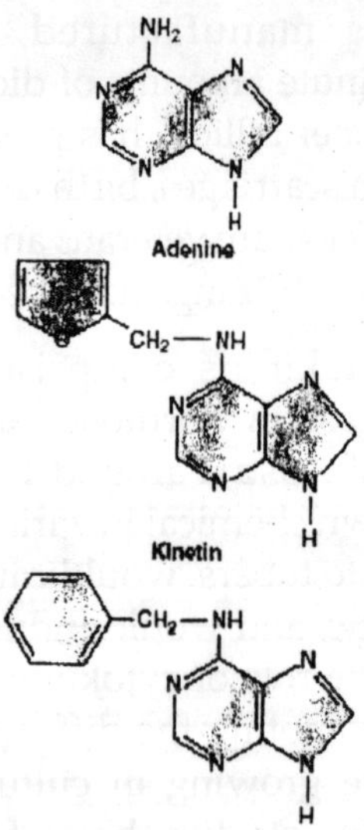

Cytokinins. Two commonly used synthetic cytokinins: kinetin and 6-benzylamino purine.

In addition, the application of cytokinins to leaves detached from a plant retards their yellowing. They function as anti-aging hormones. The action of cytokinins, like that of other hormones, has been studied in terms of its effects on growth and differentiation of masses of tissue growing in defined media. Plant tissue can form shoots, roots, or an undifferentiated mass of tissues, depending on the relative amounts of auxin and cytokinin.

In the early cell-growth experiments coconut "milk" was an essential factor. Eventually, it was discovered that coconut "milk" is not only rich in amino acids and other reduced nitrogen compounds required for growth, but it also contains cytokinins. Cytokinins seem to be essential for mitosis and cell division. They apparently promote the synthesis or activation of proteins that are specifically required for mitosis. Cytokinins have also been used against plants by pathogens.

The bacteria Agrobacterium, for example, introduces genes into the plant genome that increase the rate of cytokinin, as well as auxin, production. This causes massive cell division and the formation of a tumour called crown gall. How these hormone biosynthesis genes ended up in a bacterium is an intriguing evolutionary question. Co-evolution does not always work to the plant's advantage.

Gibberellins : Gibberellins are named after the fungus Gibberella fujikuroi, which causes rice plants, on which it is parasitic, to grow abnormally tall. Japanese plant pathologist Eiichi Kurosawa investigated Bakane ("foolish seedling") disease in the 1920s. He grew Gibberella in culture and obtained a substance that, when applied to rice plants, produced bakane. This substance was isolated and the structural formula identified by Japanese chemists in 1939. British chemists reconfirmed the formula in 1954.

Although such chemicals were first thought to be only a curiosity, they have since turned out to belong to a large class of more than 100 naturally occurring plant hormones called gibberellins. All are acidic and are usually abbreviated to GA (for gibberellic acid), with a different subscript (GA_1, GA_2, and so forth) to distinguish each one. While gibberellins function endogenously as hormones, they also function as pheromones in ferns. In ferns gibberellin-like compounds released from one gametophyte can trigger the development of male reproductive structures on a neighbouring gametophyte.

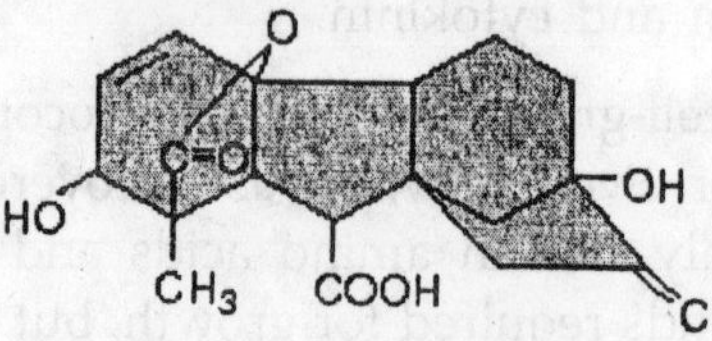

Gibberellins

Gibberellins, which are synthesised in the apical portions of stems and roots, have important effects on stem elongation. The elongation effect is enhanced if auxin is also present. The application of gibberellins to dwarf mutants is known to restore the normal growth and development in many plants. Some dwarf mutants produce insufficient amounts of gibberellin; while others lack the ability to perceive gibberellin.

The large number of gibberellins are all part of a complex biosynthetic pathway that has been unravelled using gibberellin-deficient mutants in maize (corn). While many of these gibberellins are intermediate forms in the production of GA_1, recent work shows that different forms may have specific biological roles. The role gibberellins stimulate the production of a-amylase and other hydrolytic enzymes needed for utilisation of food resources during germination and establishment of cėreal seedlings. Experimental studies in the aleurone layer surrounding the endosperms of cereal grains have shown that transcription occurs when the gibberellins initiate a burst of messenger RNA (mRNA) and protein synthesis. GA somehow enhances DNA binding proteins, which in turn allow DNA transcription of a gene. Synthesis of DNA does not seem to occur during the early stages of seed germination but becomes important when the radicle has grown through the seed coats. Gibberellins also affect a number of other aspects of plant growth and development. These hormones also hasten seed germination, apparently because they can substitute for the effects of cold or light requirements in this process. Gibberellins are used commercially to space grape flowers by extending internode length so the fruits have more room to grow.

Brassinosteroids : Although we've known about brassinosteroids for 30 years, it is only recently that they have claimed their place as a class of plant hormones. They were first discovered in Brassica pollen, hence the name. Their historical absence in discussions of hormones may be partially due to their functional overlap with other plant hormones, especially auxins and gibberellins. Additive effects among these three classes have been reported.

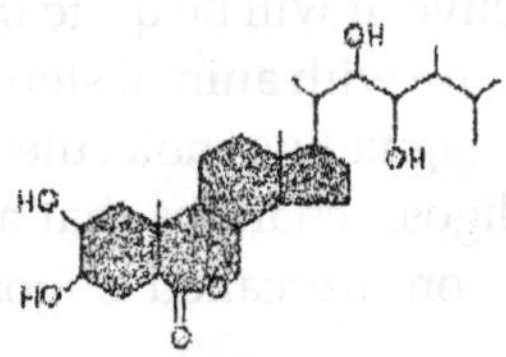

Brassinolide

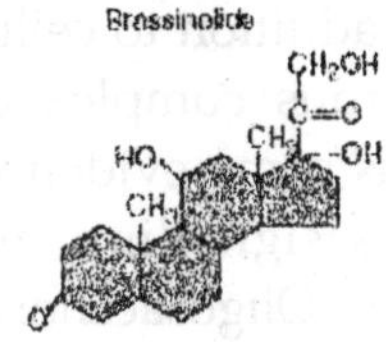

Cortisol

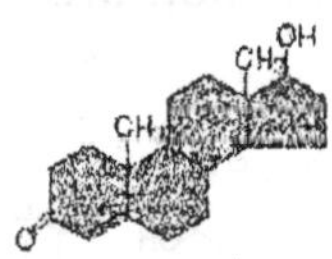

Testosterone

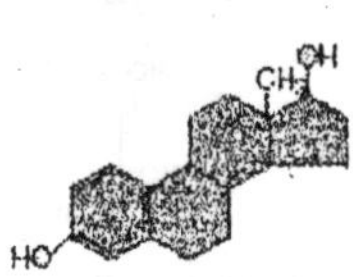

Estradiol

Brassinosteroids.

What is particularly intriguing about brassinosteroids are similarities to animal steroid hormones. One of the genes coding for an enzyme in the brassinosteroid biosynthetic pathway has significant similarity to an enzyme used in the synthesis of testosterone and related steroids. Brassinosteroids have been identified in algae and appear to be quite ubiquitous among the plants. It is plausible that their evolutionary origin predated the plant-animal split.

Brassinosteroids have a broad spectrum of physiological effects—elongation, cell division, bending of stems, vascular tissue development, delayed senescence, membrane polarisation, and reproductive development. Environmental signals can trigger brassinosteroid actions. Mutants have been identified that alter the response to brassinosteroid, but signal transaction pathways remain to be uncovered. From an evolutionary perspective, it will be quite interesting to see how these pathways compare with animal steroid signal transaction pathways function as signalling molecules as well as structural wall components. Oligosaccharides that are proposed to have a hormone-like function are called oligosaccharins.

Oligosaccharins : In addition to cellulose, plant cell walls are composed of numerous complex carbohydrates called oligosaccharides. There is some evidence that these cell wall components function as signalling molecules as well as structural wall components. Oligosacchorides that are proposed to have as hormone-like function are called oligosaccharins.

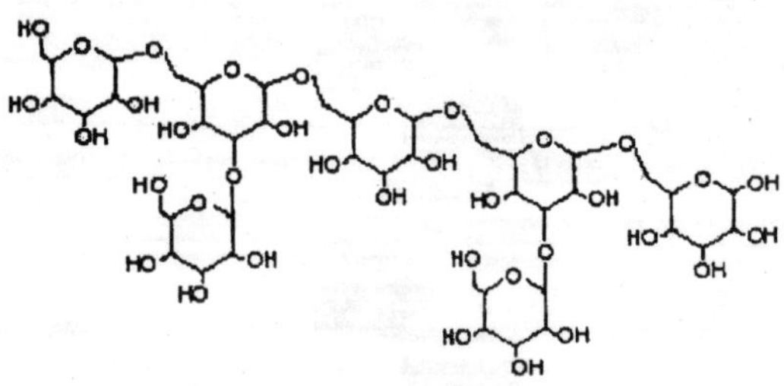

Figure: Oligosaccharins

Oligosaccharins can be released from the cell wall by

enzymes secreted by pathogens. These carbohydrates are believed to signal defence responses. Another oligosaccharin has been shown to inhibit auxin-stimulated elongation of pea stems.

These molecules are active at concentrations one to two orders of magnitude less than the traditional plant hormones. You have seen how auxin and cytokinin ratios can affect organogenesis in culture. Oligosaccharins also affect the phenotype of regenerated tobacco tissue, inhibiting root formation and stimulating flower production in tissues that are competent to regenerate flowers. How the culture results translate to in vivo systems is an open question.

The structural biochemistry of oligosaccharins makes them particularly challenging molecules to study. How they interface with cells and initiate signal transaction pathways is an open question.

Ethylene : Long before its role as a plant hormone was appreciated, the simple, gaseous hydrocarbon ethylene (H_2C—CH_2) was known to defoliate plants when it leaked from gaslights in streetlamps. Ethylene is, however, a natural product of plant metabolism that, in minute amounts, interacts with other plant hormones. When auxin is transported down from the apical meristem of the stem, it stimulates the production of ethylene in the tissues around the lateral buds and thus retards their growth.

Ethylene also suppresses stem and root elongation, probably in a similar way. An ethylene receptor has been identified and characterised. It appears to have evolved early in the evolution of Photosynthetic organisms, sharing features with environmental-sensing proteins identified in bacteria. Ethylene plays a major role in fruit ripening. At first, auxin, which is produced in significant amounts in pollinated flowers and developing fruits, stimulates ethylene production; this, in turn, hastens fruit ripening.

Complex carbohydrates are broken down into simple

sugars, chlorophylls are broken down, cell walls become soft, and the volatile compounds associated with flavour and scent in ripe fruits are produced. One of the first observations that led to the recognition of ethylene as a plant hormone was the pre-mature ripening in bananas produced by gases coming from oranges. Such relationships have led to major commercial uses of ethylene.

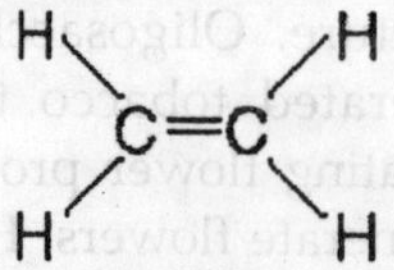

Ethylene

For example, tomatoes are often picked green and artificially ripened later by the application of ethylene. Ethylene is widely used to speed the ripening of lemons and oranges as well. Carbon dioxide has the opposite effect of arresting ripening. Fruits are often shipped in an atmosphere of carbon dioxide.

A biotechnology solution has also been developed. One of the genes necessary for ethylene biosynthesis has been cloned, and its anti-sense copy has been inserted into the tomato genome. The anti-sense copy of the gene is a nucleotide sequence that is complementary to the sense copy of the gene. In this transgenic plant, both the sense and anti-sense sequences for the ethylene biosynthesis gene are transcribed.

The sense and anti-sense mRNA sequences then pair with each other. This blocks translation, which requires single-stranded RNA; ethylene is not synthesised, and the transgenic tomatoes do not ripen. Sturdy green tomatoes can be shipped without ripening and rotting. Exposing these tomatoes to ethylene later will allow them to ripen. Studies have shown that ethylene plays an important ecological role.

Ethylene production increases rapidly when a plant is exposed to ozone and other toxic chemicals, temperature extremes, drought, attack by pathogens or herbivores, and other stresses. The increased production of ethylene that occurs can accelerate the loss of leaves or fruits that have been damaged by these stresses.

Some of the damage associated with exposure to ozone is due to the ethylene produced by the plants. The production of ethylene by plants attacked by herbivores or infected with diseases may be a signal to activate the defence mechanisms of the plants. This may include the production of molecules toxic to the pests.

Abscisic Acid : Abscisic acid, a naturally occurring plant hormone, appears to be synthesised mainly in mature green leaves, fruits, and root caps. The hormone earned its name because applications of it appear to stimulate leaf senescence (aging) and abscission, but there is little evidence that it plays an important role in this process. In fact, it is believed that abscisic acid may cause ethylene synthesis, and that it is actually the ethylene that promotes senescence and abscision.

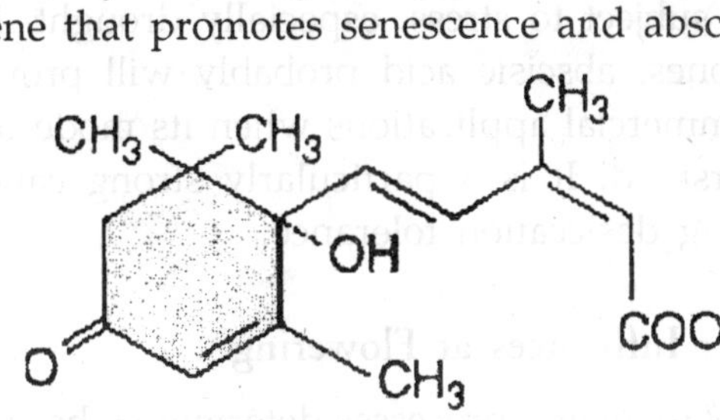

Abscisic acid

When abscisic acid is applied to a green leaf, the areas of contact turn yellow. Thus, abscisic acid has the exact opposite effect on a leaf from that of the cytokinins; a yellowing leaf will remain green in an area where cytokinins are applied. Abscisic acid probably induces the formation of winter buds—dormant buds that remain through the winter—by suppressing growth. The conversion of leaf primordia into bud scales follows.

Like ethylene, it may also suppress growth of dormant lateral buds. It appears that abscisic acid, by suppressing growth and elongation of buds, can counteract some of the effects of gibberellins (which stimulate growth and elongation of buds); it also promotes senescence by counteracting auxin (which tends to retard senescence). Abscisic acid plays a role in seed dormancy and is antagonistic to gibberellins during germination. It is also important in controlling the opening and closing of stomata.

Abscisic acid occurs in all groups of plants and apparently has been functioning as a growth-regulating substance since early in the evolution of the plant kingdom. Relatively little is known about the exact nature of its physiological and biochemical effects. These effects are very rapid—often taking place within a minute or two—and therefore they must be at least partly independent of gene expression.

Some longer-term effects of abscisic acid involve the regulation of gene expression, but the way this occurs is poorly understood. Abscisic acid levels become greatly elevated when the plant is subject to stress, especially drought. Like other plant hormones, abscisic acid probably will prove to have valuable commercial applications when its mode of action is better understood. It is a particularly strong candidate for understanding desiccation tolerance.

Environment Influences at Flowering

Carefully regulated processes determine when and where flowers will form. Plants must often gain competence to respond to internal or external signals regulating flowering. Once plants are competent to reproduce, a combination of factors including light, temperature, and both promotive and inhibitory internal signals determine when a flower is produced.

These signals turn on genes that specify where the floral organs, sepals, petals, stamens, and carpels will form. Once cells have instructions to become a specific floral organ, yet another developmental cascade leads to the three-dimensional construction of flower parts. Plants go through developmental changes leading to reproductive maturity just like many animals. This shift from juvenile to adult development is seen in the metamorphosis of a tadpole to an adult frog or caterpillar to a butterfly that can then reproduce.

Plants undergo a similar metamorphosis that leads to the production of a flower. Unlike the frog that loses its tail, plants just keep adding on structures to existing structures with their meristems. At germination, most plants are incapable of producing a flower, even if all the environmental cues are optimal. Internal developmental changes allow plants to obtain competence to respond to external and/or internal signals that trigger flower formation.

This transition is referred to as phase change. Phase change can be morphologically obvious or very subtle. Take a look at an oak tree in the winter. The lower leaves will still be clinging to the branches, while the upper ones will be gone. Those lower branches were initiated by a juvenile meristem. The fact that they did not respond to environmental cues and drop their leaves indicates that they are young branches and have not made a phase change. Ivy also has distinctive juvenile and adult phases of growth.

Stem tissue produced by a juvenile meristem initiates adventitious roots that can cling to walls. If you look at very old brick buildings that are covered with ivy, you will notice the uppermost branches are falling off because they have transitioned to the adult phase of growth and have lost the ability to produce adventitious roots. It is important to remember that even though a plant has reached the adult stage of development, it may or may not produce reproductive structures. Other factors may be necessary to trigger flowering.

Generally it is easier to get a plant to revert from an adult to vegetative state than to induce phase change experimentally. Applications of gibberellins and severe pruning can cause reversion. There is evidence in peas and Arabidopsis for genetically controlled repression of flowering. The embryonic flower mutant of Arabidopsis flowers almost immediately, which is consistent with the hypothesis that the wild-type allele suppresses flowering. It is possible that flowering is the default state and that mechanisms have evolved to delay flowering. This delay allows the plant to store more energy to be allocated for reproduction.

The best example of inducing the juvenile to adult transition comes from the construction of transgenic plants that overexpress a gene necessary for flowering, that is found in many species. This gene, LEAFY, was cloned in Arabidopsis and its promoter was replaced with a viral promoter that results in constant, high levels of LEAFY transcription.

This gene construct was then introduced into cultured aspen cells which were used to regenerate plants. When LEAFY is over-expressed in aspen, flowering occurs in weeks instead of years. Phase change requires both sufficient signal and the ability to perceive the signal. Some plants acquire competence in the shoot to perceive a signal of a certain intensity. Others acquire competence to produce sufficient promotive signal(s) and /or decrease inhibitory signal(s).

Leading to Flower Production

The environment can promote or repress flowering. In some cases, it can be relatively neutral. Light can be a signal that long, summer days have arrived in a temperate climate and conditions are favourable for reproduction. In other cases, plants depend on light to accumulate sufficient amounts of sucrose to fuel reproduction, but flower independently of the length of day.

Temperature can also be used as a clue. Gibberellins are important and have been linked to the vernalisation pathway. Clearly, reproductive success would be unlikely in the middle of a blizzard. Assuming regulation of reproduction first arose in more constant tropical environments, many of the day length and temperature controls would have evolved as plants colonised more temperate climates. Plants can rely primarily on one pathway, but all three pathways can be present. The complexity of the flowering pathways has been dissected physiologically. Now analysis of flowering mutants is providing insight into the molecular mechanisms of the floral pathways. The redundancy of pathways to flowering ensures that there will be another generation.

Light-Dependent : Flowering requires much energy accumulated via photosynthesis. Thus, all plants require light for flowering, but this is distinct from the photoperiodic, or light-dependent, flowering pathway. Aspects of growth and development in most plants are keyed to changes in the proportion of light to dark in the daily 24-hour cycle (day length).

This provides a mechanism for organisms to respond to seasonal changes in the relative length of day and night. Day length changes with the seasons; the farther from the equator, the greater the variation. Flowering responses of plants to day length fall into several basic categories. When the daylight becomes shorter than a critical length, flowering is initiated in short-day plants. When the daylight becomes longer than a critical length, flowering is initiated in long-day plants.

Other plants, such as snapdragons, roses, and many native to the tropics (for example, tomatoes), will flower when mature regardless of day length, as long as they have received enough light for normal growth. These are referred to as day-neutral plants. Several grasses (for example, Indian grass, Sorghastrum nutans), as well as ivy, have two critical photoperiods; they will not flower if the days are too long, and they also will not flower if the days are too short.

In some species, there is a sharp distinction between long and short days. In others, flowering occurs more rapidly or slowly depending on the length of day. These plants rely on other flowering pathways as well and are called facultative long- or short-day plants.

The garden pea is an example of a facultative long-day

plant. In all of these plants, it is actually the length of darkness (night), not the length of day, that is physiologically significant. Using light as a cue permits plants to flower when abiotic environmental conditions are optimal, pollinators are available, and competition for resources with other plants may be less. For example, the spring ephemerals flower in the woods before the canopy leafs out, blocking sunlight necessary for photosynthesis.

At middle latitudes, most long-day plants flower in the spring and early summer; examples of such plants include clover, irises, lettuce, spinach, and hollyhocks. Short-day plants usually flower in late summer and fall, and include chrysanthemums, goldenrods, poinsettias, soybeans, and many weeds. Commercial plant growers use these responses to day length to bring plants into flower at specific times.

For example, photoperiod is manipulated in greenhouses so poinsettias flower just in time for the winter holidays. The geographic distribution of certain plants may be determined by flowering responses to day length. Photoperiod is perceived by several different forms of phytochrome and also a blue-light-sensitive molecule (cryptochrome). The conformational change in a light receptor molecule triggers a cascade of events that leads to the production of a flower.

There is a link between light and the circadian rhythm regulated by an internal clock that facilitates or inhibits flowering. At a molecular level the gaps between light signalling and production of flowers are rapidly filling in and the control mechanisms are quite complex. Here is one example of how day length affects a specific flowering gene in Arabidopsis, a facultative long-day plant that flowers in response to both far-red and blue light.

Red light inhibits flowering. The gene CONSTANS (CO) is expressed under long days but not short days. The loss of CO product does not alter when a plant flowers under short days, but delays flowering under long days. What happens is that the gene is positively regulated by cryptochrome that perceives blue light under long days.

Cryptochrome appears to inhibit the inhibition of flowering by phytochrome B exposed to red light. Simply put, flowering is promoted by repressing a gene that represses flowering. CO is a transcription factor that turns on other genes which results in the expression of LEAFY. LEAFY is one of the key genes that "tells" a meristem to switch over to flowering.

The Flowering Hormone : The Holy Grail in plant biology has been a flowering hormone, quested unsuccessfully for more than 50 years. A considerable amount of evidence demonstrates the existence of substances that promote flowering and substances that inhibit it. Grafting experiments have shown that these substances can move from leaves to shoots. The complexity of their interactions, as well as the fact that multiple chemical messengers are evidently involved, has made this scientifically and commercially interesting search very difficult, and to this day, the existence of a flowering hormone remains strictly hypothetical.

We do know that LEAFY can be expressed in the vegetative as well as the reproductive portions of plants. Clearly, information about day length gathered by leaves is transmitted to shoot apices. Given that there are multiple pathways to flowering, several signals may be facilitating communication between leaves and shoots.

Temperature-Dependent Pathway : Lysenko solved the problem of winter wheat seed rotting in the fields in Russia by chilling the seeds and planting them in the spring. Winter wheat would not flower without a period of chilling, called vernalisation. Unfortunately a great many problems, including mistreatment of Russian geneticists, resulted from this scientifically significant discovery.

Lysenko erroneously concluded that he had converted one species, winter wheat, to another, spring wheat, by simply altering the environment. There was a shift from science to politics. Genetics and Darwinian evolution were suspect for half a century. Social history aside, the valuable lesson here is that cold temperatures can accelerate or permit flowering in many species. As with light, this ensures that plants flower at more optimal times.

Vernalisation may be necessary for seeds or plants in later stages of development. Analysis of mutants in Arabidopsis and pea indicate that vernalisation is a separate flowering pathway that may be linked to the hormone gibberellin. In this pathway, repression may also lead to flowering. High levels of one of the genes in the pathway may block the promotion of flowering by gibberellins. When plants are chilled, there is less of this gene product and gibberellin activity may increase.

It is known that gibberellins enhance the expression of LEAFY. One proposal is that both the vernalisation and autonomous pathways share a common intersection affecting gibberellin promotion of flowering. Weigel has shown that gibberellin actually binds the promoter of the LEAFY gene, so its effect on flowering is direct. The connection between gibberellin levels and temperature also needs to be understood.

Autonomous Pathway : The autonomous pathway to flowering is independent of external cues except for basic nutrition. Presumably this was the first pathway to evolve. Day-neutral plants often depend primarily on the autonomous pathway which allows plants to "count" and "remember." A field of day-neutral tobacco will produce a uniform number of nodes before flowering. If the shoots of these plants are removed at different positions, axillary buds will grow out and produce the same number of nodes as the removed portion of the shoot. At a certain point in development shoots become committed or determined to flower.

The upper axillary buds of flowering tobacco will remember their position when rooted or grafted. The terminal shoot tip becomes florally determined about four nodes before it initiates a flower. In some other species, this commitment is less stable and/or occurs later. How do shoots know where they are and at some point "remember" that information? It is clear that there are inhibitory signals from the roots. If bottomless pots are continuously placed over a growing tobacco plant and filled with soil, flowering is delayed by the formation of adventitious roots.

Control experiments with leaf removal show that it is the addition of roots and not the loss of leaves that delays flowering. A balance between floral promoting and inhibiting signals may regulate when flowering occurs in the autonomous pathway and the other pathways as well. Determination for flowering is tested at the organ or whole plant level by changing the environment and ascertaining whether or not the fate has changed. How does floral determination correlate with molecular level changes?

In Arabidopsis, floral determination correlates with the increase of LEAFY gene expression and has occurred by the time a second flowering gene, APETALA1, is expressed. Because all three flowering pathways appear to converge with increased levels of LEAFY, this determination event should occur in species with a variety of balances among the pathways.

Floral Meristem and Floral Organs

Arabidopsis and snapdragon are valuable model systems for identifying flowering genes and understanding their interactions. The three pathways, lead to an adult meristem becoming a floral meristem by either activating or repressing the inhibition of floral meristem identity genes. Two of the key floral meristem identity genes are LEAFY and APETALA1. These genes establish the meristem as a flower meristem.

They then turn on floral organ identity genes. The floral organ identity genes define four concentric whorls moving inward in the floral meristem as sepal, petal, stamen, and carpel. Meyerowitz and Coen proposed a model, called the ABC model, to explain how three classes of floral organ identity genes could specify four distinct organ types. The ABC model proposes that three classes of organ identity genes (A, B, and C) specify the floral organs in the four floral whorls. By studying mutants the researchers have determined the following:

- Class A genes alone specify the sepals.
- Class A and class B genes together specify the petals.
- Class B and class C genes together specify the stamens.
- Class C genes alone specify the carpels.

The beauty of their ABC model is that it is entirely testable by making different combinations of floral organ identity mutants. Each class of genes is expressed in two whorls, yielding four different combinations of the gene products. When any one class is missing, there are aberrant floral organs in predictable positions. It is important to recognise that this is actually only the beginning of the making of a flower.

These organ identity genes are transcription factors that turn on many more genes that will actually give rise to the three-dimensional flower. There are also genes that "paint" the petals. Complex biochemical pathways lead to the accumulation of anthocyanin pigments in vacuoles. These pigments can be orange, red, or purple and the actual colour is influenced by pH and by the shape of the petal.

Formation of Gametes : The ovule within the carpel has origins more ancient than the angiosperms. Floral parts are modified leaves, and within the ovule is the female gametophyte. This next generation develops from placental tissue in the ovary. A megaspore mother cell develops and meiotically gives rise to the embryo sac. Usually two layers of integument tissue form around this embryo sac and will become the seed coat. Genes responsible for the initiation of integuments and also those responsible for the formation of the integument have been identified. Some also affect leaf structure.

It is indeed a metamorphosis, but the subtle shift from mitosis to meiosis in the megaspore mother cell leading to the development of a haploid, gamete-producing gametophyte is perhaps even more critical. The same can be said for pollen formation in the anther of the stamen. The flower houses the haploid generations that will produce gametes. The flower also functions to increase the probability that male and female gametes from different (or sometimes the same plant) will unite.

Many Short-term Responses

Larger predators, microbes, water, and wind often present a plant with rapid immediate stress. Response, to be effective, must also be immediate. There is little time for growth, and plants instead invoke a variety of other kinds of responses. Many environmental cues trigger rapid and reversible localised plant movements, for example. The rapid folding of leaves can startle a potential predator. Leaf folding can also prevent water loss by reducing the surface area available for transpiration. Some localised plant movements are triggered by unpredictable environmental signals.

Other movements are tied into daily internal rhythms established by cyclic environmental signals like light and temperature. Plants lack a nervous system in the conventional sense. Some of the rapid signalling, however, is the result of electric charge moving through an organ as a wave of membrane ion exchange, not unlike that seen in animals. This is translated into movement by changing the turgor pressure of cells.

Turgor Movement : Turgor is pressure within a living cell resulting from diffusion of water into it. If water leaves turgid cells (ones with turgor pressure), the cells may collapse, causing plant movement; conversely, water entering a limp cell may also cause movement as the cell once more becomes turgid. Many other plants, including those of the legume family (Fabaceae), exhibit leaf movements in response to touch or other stimuli. After exposure to a stimulus, the changes in leaf orientation are mostly associated with rapid turgor pressure changes in pulvini (singular: pulvinus), which are multicellular swellings located at the base of each leaf or leaflet.

When leaves with pulvini, such as those of the sensitive plant (Mimosa pudica), are stimulated by wind, heat, touch, or, in some instances, intense light, an electrical signal is generated. The electrical signal is translated into a chemical signal, with potassium ions, followed by water, migrating from the cells in one half of a pulvinus to the intercellular spaces in the other half. The loss of turgor in half of the pulvinus causes the leaf to "fold." The movements of the leaves and leaflets of the sensitive plant are especially rapid; the folding occurs within a second or two after the leaves are touched.

Over a span of about 15 to 30 minutes after the leaves and leaflets have folded, water usually diffuses back into the same cells from which it left, and the leaf returns to its original position. Some of the most familiar of these reversible changes are seen in leaves and flowers that "open" during the day and "close" at night. For example, the flowers of four o'clocks open at 4 P.M. and evening primrose petals open at night. The blades of plant leaves that exhibit such a daily shift in position may not actually fold; instead, their orientation may be changed as a result of turgor movements.

Bean leaves are horizontal during the day when their pulvini are turgid, but become more or less vertical at night as the pulvini lose turgor. These sleep movements reduce water loss from transpiration during the night, but maximise Photosynthetic surface area during the day. In these cases, the movement is closely tied to an internal rhythm.

Circadian Clocks : They have endogenous circadian clocks that set a rhythm with a period of about 24 hours (actually it is closer to 22 or 23 hours). While there are shorter and much longer, naturally occurring rhythms, circadian rhythms are particularly common and widespread because of the day-night cycle on earth. Jean de Mairan, a French astronomer, first identified circadian rhythms in 1729.

When de Mairan put the plants in total darkness, they continued "sleeping" and "waking" just as they did when exposed to night and day. This is one of four characteristics of a circadian rhythm: it must continue to run in the absence of external inputs. It must be about 24 hours in duration and can be reset or entrained. The fourth characteristic is that the clock can compensate for differences in temperature.

This is quite unique when you consider what you know about biochemical reactions; most rates of reactions vary significantly based on temperature. Circadian clocks exist in many organisms and appear to have evolved independently multiple times. The mechanism behind the clock is not fully understood, but is being actively investigated at the molecular level.

Plant Defence Responses : Interactions between plants and other organisms can be symbiotic (for example, nitrogen-fixing bacteria and mycorrhizae) or pathogenic. In evolutionary terms, these two types of interactions may simply be opposite sides of the same coin. The interactions have many common aspects and are the result of co-evolution between two species that signal and respond to each other. In the case of pathogens, the microbe or pest is "winning," at least for that second in evolutionary time.

Half a century ago, Flor proposed that there is a plant resistance gene whose product interacts with that of a pathogen avirulence gene (avr). This is called the gene-for-gene model and several pairs of avr and R genes have been cloned in different species pathogenised by microbes, fungi, and insects, in one case. This has been motivated partially by the agronomic benefit of identifying genes that can be added to protect other plants from invaders. Much is now known about the signal transaction pathways that follow the recognition of the pathogen by the R gene.

These pathways lead to the triggering of the hypersensitive response (HR) which leads to rapid cell death around the source of the invasion and also a longer-term resistance. There is not always a gene-for-gene response, but plants still have defence responses to pathogens and also mechanical wounding. Some of the response pathways may be similar. Also, oligosaccharins in the cell walls may serve as recognition and signalling molecules.

There are specific molecules released from the root hairs of the host that the parasitic plant recognises and responds to with invasive action. Less is known about the host response and so far the different defence genes that are activated appear to be ineffective. When a plant is attacked and there is gene-for-gene recognition, the HR response leads to very rapid cell death around the site of attack. This seals off the wounded tissue to prevent the pathogen or pest from moving into the rest of the plant. Hydrogen peroxide and nitric oxide are produced and may signal a cascade of biochemical events resulting in the localised death of host cells.

They may also have negative effects on the pathogen, although antioxidant abilities have coevolved in pathogens. Other anti-microbial agents produced include the phytoalexins which are chemical defence agents. A variety of pathogenesis-related genes (PR genes) are expressed and their proteins can either function as anti-microbial agents or signals for other events that protect the plant. In the case of virulent invaders (no R gene recognition), there are changes in local cell walls that at least partially block the movement of the pathogen or pest farther into the plant.

In this case there is not an HR response and the local plant cells are not suicidal. When an insect takes a mouthful of a leaf, defence responses are also triggered. Mechanical damage causes responses that have some similar components, but the reaction may be slower. Biochemically, it is distinct from some of the events triggered by signals in the insect's mouth. Such responses are collectively called wound responses. Wound responses are a challenge in designing other types of experiments with plants that involve cutting or otherwise mechanically damaging the tissue. It is important to run control experiments to be sure you are answering your question and not observing a wound response.

In addition to the HR or other local responses, plants are capable of a systemic response to a pathogen or pest attack. This is called a systemic acquired response (SAR). Several pathways lead to broad-ranging resistance that lasts for a period of days. The signals that induce SAR include salicylic acid and jasmonic acid. Salicylic acid is the active ingredient in aspirin too. SAR allows the plant to respond more quickly if it is attacked again. However, this is not the same as the human immune response where antibodies (proteins) that recognise specific antigens (foreign proteins) persist in the body. SAR is neither as specific or long lasting.

tissue. It is important to run control experiments to be sure you are answering your question and not observing a wound response.

In addition to the HR or other local responses, plants are capable of a systemic response to a pathogen or pest attack. This is called a systemic acquired response (SAR). Several pathways lead to broad-ranging resistance that lasts for a period of days. The signals that induce SAR include salicylic acid and jasmonic acid. (Salicylic acid is the active ingredient in aspirin too.) SAR allows the plant to respond more quickly if it is attacked again. However, this is not the same as the human immune response where antibodies (proteins) that neutralise specific antigens (foreign proteins) persist in the body. SAR is neither as specific or long lasting.

10

Combination of Life

All matter, from a rock to an animal to the magma at the centre of the Earth, is made from different combinations of 92 naturally occurring substances known as elements. The smallest quantity of an element that still exhibits the characteristics of that element is known as an atom. For example, one atom of carbon is the smallest piece of matter that still retains the chemical and physical characteristics of carbon. Atoms are made up of even smaller particles called electrons, protons, and neutrons. Each of these particles has a different electrical charge.

Protons are positively charged, neutrons have no charge, and electrons are negatively charged. The protons and neutrons of an atom reside in a central body called a nucleus. Electrons appear around the nucleus within orbitals of varying energy. Overall, the atom is neutrally charged with equal numbers of positively charged protons and negatively charged electrons.

Elements are distinguished by the number of protons in their nuclei. All atoms containing six protons are called carbon. Any element with one proton is called hydrogen. Only the number of protons-and not the number of neutrons or electrons-distinguishes elements from each other.

Isotopes and Ions

An atom that contains a larger or smaller number of neutrons than usual is called an isotope. Carbon usually has six protons and six neutrons and can be called carbon-12 because the number of its protons and neutrons add up to 12. But some carbon atoms have seven or even eight neutrons. These two isotopes are called carbon-13 and carbon-14. Isotopes do not have charge, because the numbers of positive and negative particles remain balanced. Even though they have different masses, isotopes of the same element all have similar chemical properties, because the number of electrons (not the number of neutrons or protons) determines the way an atom will interact with other atoms.

Ions are atoms that either lack or have extra electrons. Because these atoms have unequal numbers of electrons and protons, they are charged particles and are often quite chemically interactive with other atoms. Though the SAT II Biology Test rarely asks direct questions about ions, ions do play an important role in many biological processes and phenomena, so understanding the basics of ions can help you understand the processes that the test covers.

Molecules and Compounds

Atoms combine with each other in chemical reactions to create molecules, unique substances with physical and chemical properties distinct from those of their constituent elements. Combining two hydrogen atoms with one oxygen atom creates water, which has very different characteristics than hydrogen or oxygen do alone. Molecules such as water containing more than one type of element can also be called compounds.

A water molecule made up of oxygen and hydrogen can be called a compound; a hydrogen molecule, which contains only two hydrogen atoms, cannot be called a compound. You may have heard water referred to as H_2O. This notation is the standard way of representing molecules and compounds by shorthand. The "H" and "O" stand for the elements hydrogen and oxygen, and the subscript indicates that water contains two parts hydrogen for every one part oxygen. You can create the formula for any compound by writing down the letter symbol of each of its constituent elements and using subscripted numbers to indicate how many atoms of each element are present.

Chemical Bonds

The connections between the atoms in a compound are called chemical bonds. Atoms form bonds by sharing their electrons with each other, relying on the power of electric charge to keep themselves attached. Molecules and compounds can also bond with each other.

Covalent Bonds : Important bonds between atoms are covalent and ionic bonds. Bonds formed through the more or less equal sharing of electrons between atoms are known as covalent bonds. If the electrons in a covalent bond are shared equally, the resulting bond is called a non-polar covalent bond. When one atom pulls the shared electrons towards itself a little more tightly than the other, the resulting covalent bond is said to be a polar bond. In a polar bond, the atom that pulls electrons towards itself gains a slight negative charge (because electrons have a negative charge). Since the other atom partially loses an electron, it gains a slight positive charge. For example, the atoms in water form polar bonds because oxygen, which has eight protons in its nucleus, has a greater pull on electrons than hydrogen, which has only one proton.

Ionic Bonds : Polar covalent bonds involve the unequal sharing of electrons. This inequality is brought to an extreme in a bonding arrangement called an ionic bond. In an ionic bond, one atom pulls the shared electrons away from the other atom entirely. Ionic bonds are stronger than polar bonds. One example of ionic bonding is the reaction between sodium (Na) and chlorine (Cl) to form table salt (NaCl). The chlorine atom steals an electron from the sodium atom. Because it loses an electron, the sodium atom develops a charge of +1. The chlorine atom has a charge of -1, since it gained an electron.

Dipole-Dipole Bonds : Bonds between molecules or compounds are called dipole-dipole bonds. As seen in polar covalent compounds, due to the unequal sharing of electrons, some molecules have a slightly positive and a slightly negative end to them, or a dipole (di-pole = two magnetic poles). These compounds can form weak bonds with one another without combining together completely to create new compounds. This type of bonding, known as dipole-dipole interaction, takes places when the positively charged end of one polar covalent compound (d+) comes in contact with the negatively charged end of another polar covalent compound (d-):

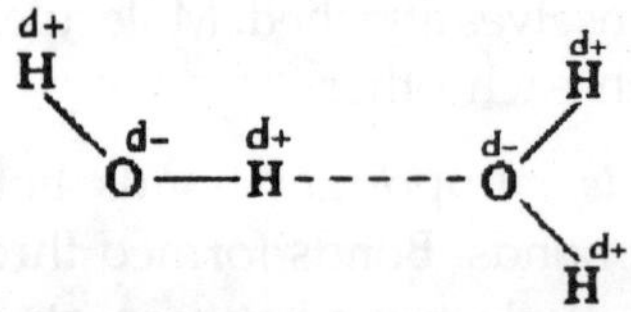

Dipole-dipole interactions are much weaker than the bonds within molecules, but they play a very important role in the chemistry of life. Perhaps the most important dipole-dipole bond in biochemistry is the dipole-dipole interaction between positively charged hydrogen molecules and negatively charged oxygen molecules. This reaction is so important, it gets its own special name: hydrogen bond. These bonds account for many of the exceptional properties of water and have important effects on the structure of proteins and DNA.

Acids and Bases

Sometimes atoms give their electrons up altogether instead of sharing them in a chemical bond. This process is known as disassociation. Water, for instance, dissociates by the following formula:

$H_2O \rightarrow H^+ + OH$

The hydrogen atom gives up a negatively charged electron, gaining a positive charge, and the OH compound gains a negatively charged electron, taking on a negative charge. The H^+ is known as a hydrogen ion and OH^- ion is known as a hydroxide ion. The disassociation of water produces equal amounts of hydrogen and hydroxide ions.

However, the disassociation of some compounds produces solutions with high proportions of either hydrogen or hydroxide ions. Solutions high in hydrogen ions are known as acids, while solutions high in hydroxide ions are known as bases. Both types of solution are extremely reactive-likely to form bonds-because they contain so many charged particles. The technical definition of an acid is that it is a hydrogen ion donor, or a proton donor, as hydrogen ions are consist of only a single proton. Acids put H^+ ions into solution. The definition of a base is a little more complicated: they are H^+ ion or proton acceptors, which means that they remove H^+ ions from solution. Some bases can directly produce OH^- ions that will take H^+ out of solution. NaOH is an example of this type of base:

$$NaOH \rightarrow Na^+ + OH—$$

A second type of base can directly take H^+ out of an H_2O solution. Ammonia (NH3) is a common example of this sort of base:

$$NH_3 + H_2O \rightarrow NH_4^+ + OH^-$$

From time to time, the SAT II Biology has been known to ask whether ammonia is a base.

pH Scale : The pH scale, which ranges from 0 to 14, measures the degree to which a solution is acidic or basic. If the proportion of hydrogen ions in a solution is the same as the proportion of hydroxide ions or equivalent, the solution has a pH of 7, which is neutral. The most acidic solutions (those with a high proportion of H^+) have pHs approaching 0, while the most basic solutions (those with a high proportion of OH^- or equivalent) have pHs closer to 14.

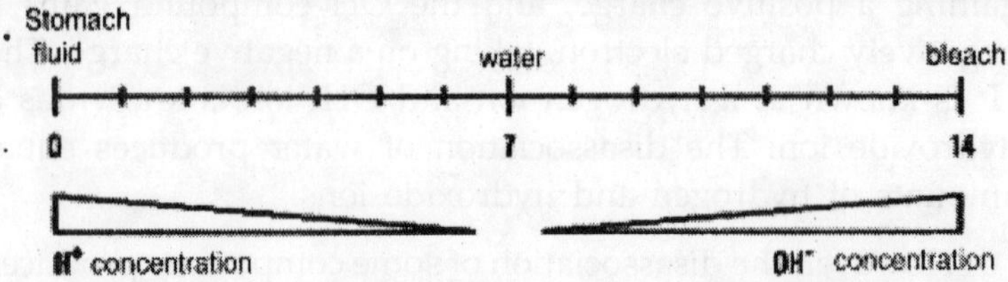

Water has a pH of 7 because it has equal proportions of H^+ and OH^- ions. In contrast, when a compound called hydrogen fluoride (HF) disassociates, it forms only hydroxide ions. HF is therefore quite acidic and has a pH well below 7. Some acids are more acidic than others because they put more H^+ ions into solution. Stomach fluid, for example, is more acidic than saliva. When sodium hydroxide (NaOH) disassociates, it forms only hydroxide ions, making it a base and giving it a pH above 7. Like acids, bases can be strong or weak depending on how many hydroxide ions they put in solution or how many hydrogen ions they take out of solution.

Buffers : Some substances resist changes in pH even when acids or bases are added to them. These substances are known as buffers. The cell contains many buffers because wide swings in pH can negatively impact the chemical reactions of cell processes.

Life and Central Elements

Of the 92 naturally existing elements on the Earth, only 25 play a role in the chemical processes of life. Of these 25, four elements constitute more than 98 per cent of all biological matter: carbon (C), oxygen (O), hydrogen (H), and nitrogen (N). Virtually every important organic compound is made up of these four elements. The Big 4 of organic elements can be cut down even further to a Supreme 1: carbon is the most important biological molecule, both for life as we know it and on the SAT II.

Carbon : Carbon is the central element of life. Its important role stems from its ability to form four chemical bonds with other elements at the same time:

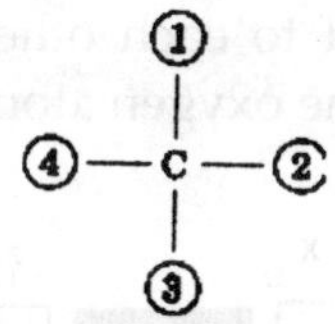

Carbons often attach to other carbon atoms, forming long chains called hydrocarbons. These molecules get their name because the central carbons also bond to hydrogen:

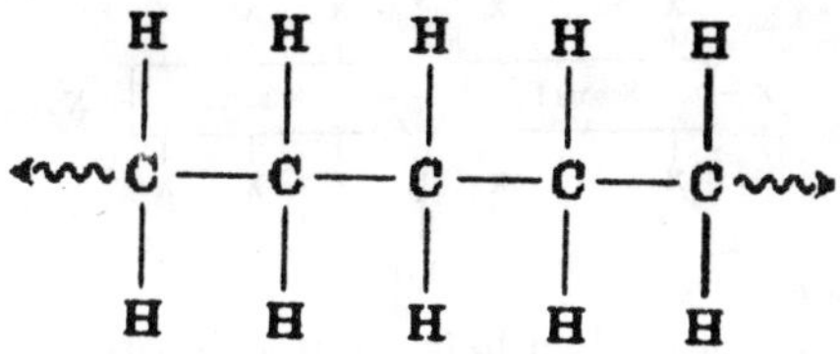

In addition to making a connection to four other atoms, carbon also has the ability to make two or three separate connections with the same single partner (and make its remaining one or two bonds with other substances). These bonds, which are stronger than single bonds, are known as double or triple bonds, respectively.

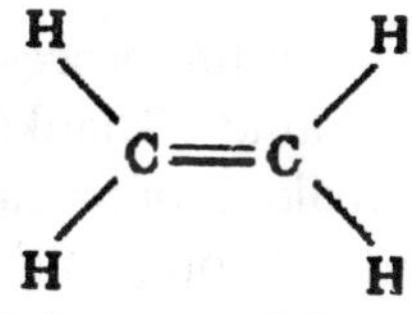

Monomers and Polymers : Many biological molecules consist of basic units that are strung together to form long chains, much like beads are placed on a string to make a necklace. There can be some variation in these basic units, which are known as monomers. Two monomers connected to each other are known as a dimer; a chain of monomers is called a polymer.

Polymers can be formed by many different types of chemical reactions. One special reaction, however, is particularly important in producing the polymers found in the chemistry of life. This reaction involves a carbon that has a hydrogen atom attached and a carbon that has an OH^- group attached.

When the carbons bond to each other, they release a water molecule formed from the oxygen atom and the two hydrogen atoms.

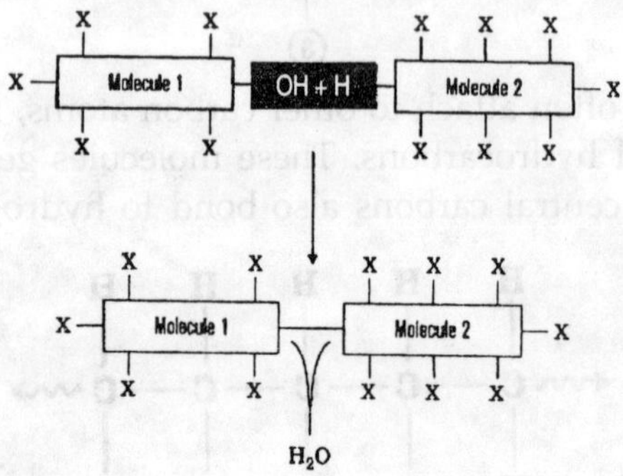

Because a water molecule is created in order to join the two monomers, this reaction is known as dehydration synthesis. The reverse of dehydration synthesis, when a water molecule is inserted into a polymer to break off a monomer, is called hydrolysis.

Life's Molecules

The elements involved in life processes can, and do, form millions of different compounds. Thankfully, these millions of compounds fall into four major groups: carbohydrates, proteins, lipids, and nucleic acids. Though all of these groups are organised around carbon, each group has its own special structure and function.

Carbohydrates : Carbohydrates are compounds that have carbon, hydrogen, and oxygen atoms in a ratio of about 1:2:1. If you're stuck on an SAT II Biology question about whether a compound is a carbohydrate, just count up the atoms and see if they fit this ratio. Carbohydrates are often sugars, which provide energy for cellular processes. Like all of the biologically important classes of compounds, carbohydrates can be monomers, dimers, or polymers. The names of most carbohydrates end in "-ose": glucose, fructose, sucrose, and maltose are some common examples.

Monosaccharides : Carbohydrate monomers are known as monosaccharides. This group includes glucose, $C_6H_{12}O_6$, which is a key substance in biochemistry. Sugars that an animal eats are converted into glucose, which is then converted into energy to fuel the animal's activities by respiration. Glucose has a cousin called fructose with the same chemical formula.

Glucose and fructose differ in one important way: glucose has a double-bonded oxygen on the top carbon, while fructose has its double-bonded carbon on the second carbon. This difference is most apparent when the two monosaccharides are in their ring forms.

Glucose generally forms a hexagonal ring (six sided), while fructose forms a pentagonal ring (five sided). Whereas fructose is the sugar most often found in fruits, glucose is most often used as the major source of energy for cellular activities.

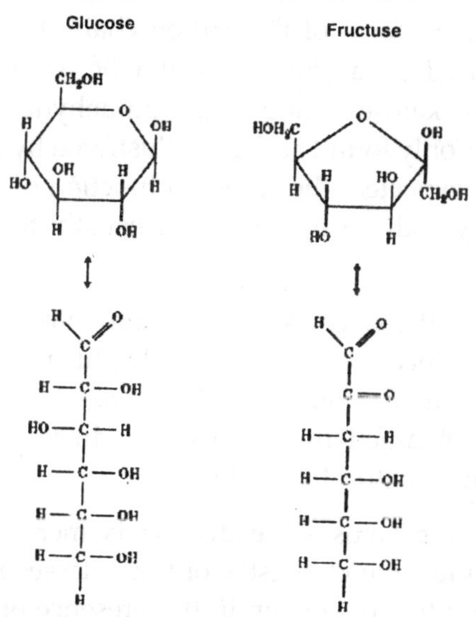

Properties of Glucose : Carbohydrates have been given non-systematic names, although the suffix ose is generally used. The most common carbohydrate is glucose ($C_6H_{12}O_6$).

Applying the terms defined above, glucose is a monosaccharide, an aldohexose (note that the function and size classifications are combined in one word) and a reducing sugar.

The general structure of glucose and many other aldohexoses was established by simple chemical reactions. The following diagram illustrates the kind of evidence considered, although some of the reagents shown here are different from those used by the original scientists.

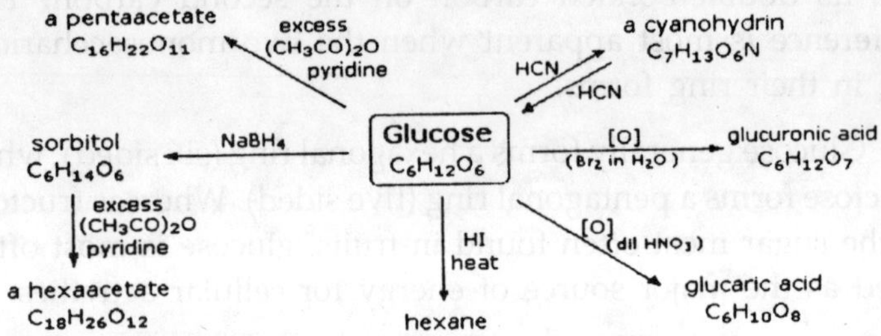

Ketoses : If a monosaccharide has a carbonyl function on one of the inner atoms of the carbon chain it is classified as a Ketose. Dihydroxyacetone may not be a sugar, but it is included as the ketose analog of glyceraldehyde. The carbonyl group is commonly found at C-2, as illustrated by the following examples. As expected, the carbonyl function of a ketose may be reduced by sodium borohydride, usually to a mixture of epimeric products.

D-Fructose, the sweetest of the common natural sugars, is for example reduced to a mixture of D-glucitol (sorbitol) and D-mannitol, named after the aldohexoses from which they may also be obtained by analogous reduction. Mannitol is itself a common natural carbohydrate.

Although the ketoses are distinct isomers of the aldose monosaccharides, the chemistry of both classes is linked due to their facile inter-conversion in the presence of acid or base catalysts. This inter-conversion, and the corresponding epimerisation at sites alpha to the carbonyl functions, occurs by way of an enediol tautomeric intermediate.

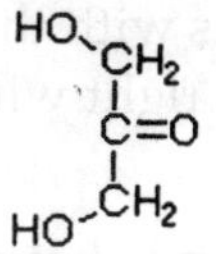

dihydroxyacetone

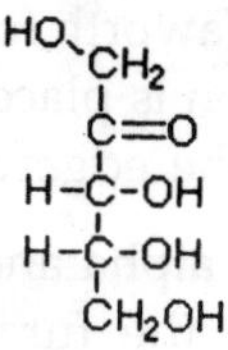

D-(-)-ribulose

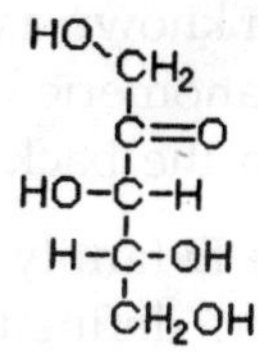

D-(-)-xylulose

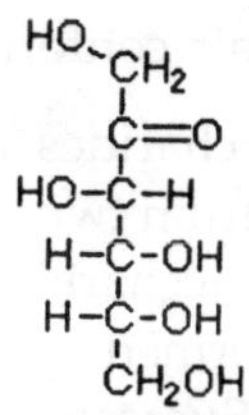

D-(-)-fructose

Because of base-catalysed isomerisations of this kind, the Tollens' reagent is not useful for distinguishing aldoses from ketoses or for specific oxidation of aldoses to the corresponding aldonic acids. Oxidation by HOBr is preferred for the latter conversion.

Cyclic Forms of Monosaccharides : The preferred structural form of many monosaccharides may be that of a cyclic hemiacetal. Five and six-membered rings are favoured over other ring sizes because of their low angle and eclipsing strain. Cyclic structures of this kind are termed furanose (five-membered) or pyranose (six-membered), reflecting the ring size relationship to the common heterocyclic compounds furan and pyran shown on the right. Ribose, an important aldopentose, commonly adopts a furanose structure, as shown in the following illustration.

By convention for the D-family, the five-membered furanose ring is drawn in an edgewise projection with the ring oxygen positioned away from the viewer. The anomeric carbon atom is placed on the right. The upper bond to this carbon is defined as beta, the lower bond then is alpha. The cyclic pyranose forms of various monosaccharides are often drawn in a flat projection known as a Haworth formula. As with the furanose ring, the anomeric carbon is placed on the right with the ring oxygen to the back of the edgewise view.

In the D-family, the alpha and beta bonds have the same orientation defined for the furanose ring. These Haworth formulas are convenient for displaying stereochemical relationships, but do not represent the true shape of the molecules. We know that these molecules are actually puckered in a fashion we call a chair conformation.

Disaccharides : Disaccharides are carbohydrate dimers. These dimers are formed from two monomers by dehydration synthesis. Any two monosaccharides can form a disaccharide. For example, maltose is formed by the dehydration synthesis of two glucose molecules. Sucrose, common table sugar, comes from the linkage of one molecule of glucose and one of fructose.

The glycoside bond may be alpha, as in maltose and trehalose, or beta as in cellobiose and gentiobiose. Acid-catalysed hydrolysis of these disaccharides yields glucose as the only product. Enzyme-catalysed hydrolysis is selective for a specific glycoside bond, so an alpha-glycosidase cleaves maltose and trehalose to glucose, but does not cleave cellobiose or gentiobiose. A beta-glycosidase has the opposite activity. In order to draw a representative structure for cellobiose, one of the glucopyranose rings must be rotated by 180°, but this feature is often omitted in favour of retaining the usual perspective for the individual rings.

This leaves the anomeric carbon free, so cellobiose and maltose both may assume alpha and beta anomers at that site. Gentiobiose has a beta-glycoside link, originating at C-1 and terminating at C-6. Because cellobiose, maltose and gentiobiose are hemiacetals they are all reducing sugars. Trehalose, a disaccharide found in certain mushrooms, is a bis-acetal, and is therefore a non-reducing sugar.

A systematic nomenclature for disaccharides exists, but as the following examples illustrate, these are often lengthy.

Cellobiose	:	4-O-β-D-Glucopyranosyl-D-glucose (the beta-anomer is drawn)
Maltose	:	4-O-α-D-Glucopyranosyl-D-glucose (the beta-anomer is drawn)
Gentiobiose	:	6-O-β-D-Glucopyranosyl-D-glucose (the alpha-anomer is drawn)
Trehalose	:	α-D-Glucopyranosyl-α-D-glucopyranoside

Although all the disaccharides shown here are made up of two glucopyranose rings, their properties differ in interesting ways. Maltose, sometimes called malt sugar, comes from the hydrolysis of starch. It is about one third as sweet as cane sugar (sucrose), is easily digested by humans, and is fermented by yeast.

Cellobiose is obtained by the hydrolysis of cellulose. It has virtually no taste, is indigestible by humans, and is not fermented by yeast. Some bacteria have beta-glucosidase enzymes that hydrolyse the glycosidic bonds in cellobiose and cellulose. The presence of such bacteria in the digestive tracts of cows and termites permits these animals to use cellulose as a food. Finally, it may be noted that trehalose has a distinctly sweet taste, but gentiobiose is bitter.

Disaccharides made up of other sugars are known, but glucose is often one of the components. Lactose, also known as milk sugar, is a galactose-glucose compound joined as a beta-glycoside. It is a reducing sugar because of the hemiacetal function remaining in the glucose moiety. Many adults, particularly those from regions where milk is not a dietary staple, have a metabolic intolerance for lactose. Infants have a digestive enzyme which cleaves the betaglycoside bond in lactose, but production of this enzyme stops with weaning. Cheese is less subject to the lactose intolerance problem, since most of the lactose is removed with the whey.

Sucrose, or cane sugar, is our most commonly used sweetening agent. It is a non-reducing disaccharide composed of glucose and fructose joined at the anomeric carbon of each by glycoside bonds (one alpha and one beta). In the formula shown here the fructose ring has been rotated 180° from its conventional perspective.

Polysaccharides : Polysaccharides can consist of as few as three and as many as several thousand monosaccharides. Depending on their structure and the monosaccharides they contain, polysaccharides can function as a means of storing excess energy or provide structural support. When cells ingest more carbohydrates than they need for fuel, they link the sugars together to form polysaccharides. The structure of these polysaccharides is different in plants and animals: in plants, polysaccharides take the form of starch, whereas in animals, they are linked in a structure called glycogen.

Polysaccharides can also have structural roles in plants and animals. Cellulose, which forms the cell walls of plant cells, is a structural polysaccharide. In animals, the polysaccharide chitin forms the hard outer armour of insects, crabs, spiders, and other arthropods. Many fungi also use chitin as a structural carbohydrate. Polysaccharides are sometimes called glycans. The most important compounds in this class, cellulose, starch and glycogen are all polymers of glucose. This is easily demonstrated by acid-catalysed hydrolysis to the monosaccharide. Since partial hydrolysis of cellulose gives varying amounts of cellobiose, we conclude the glucose units in this macromolecule are joined by beta-glycoside bonds between C-1 and C-4 sites of adjacent sugars.

Partial hydrolysis of starch and glycogen produces the disaccharide maltose together with low molecular weight dextrans, polysaccharides in which glucose molecules are joined by alpha-glycoside links between C-1 and C-6, as well as the alpha C-1 to C-4 links found in maltose. Over half of the total organic carbon in the earth's biosphere is in cellulose. Cotton fibres are essentially pure cellulose, and the wood of bushes and trees is about 50 per cent cellulose. As a polymer of glucose, cellulose has the formula $(C_6H_{10}O_5)n$ where n ranges from 500 to 5,000, depending on the source of the polymer.

The glucose units in cellulose are linked in a linear fashion. The beta-glycoside bonds permit these chains to stretch out, and this conformation is stabilised by intramolecular hydrogen bonds. A parallel orientation of adjacent chains is also favoured by intermolecular hydrogen bonds. Although an individual hydrogen bond is relatively weak, many such bonds acting together can impart great stability to certain conformations of large molecules. Most animals cannot digest cellulose as a food, and in the diets of humans this part of our vegetable intake functions as roughage and is eliminated largely unchanged. Some animals (the cow and termites, for example) harbour intestinal microorganisms that breakdown cellulose into monosaccharide nutrients by the use of beta-glycosidase enzymes.

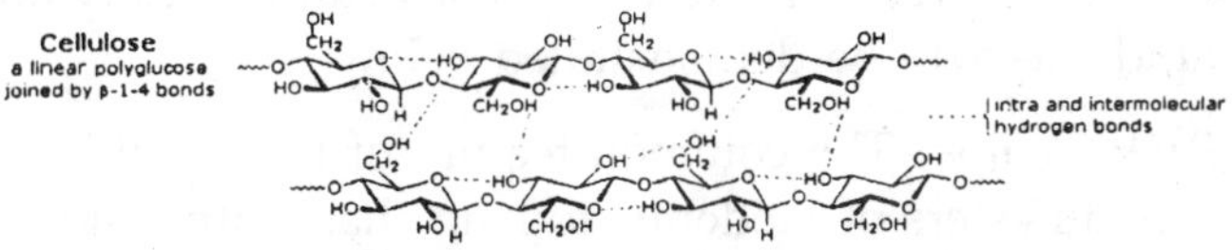

Starch is a polymer of glucose, found in roots, rhizomes, seeds, stems, tubers and corms of plants, as microscopic granules having characteristic shapes and sizes. Most animals, including humans, depend on these plant starches for nourishment. The structure of starch is more complex than that of cellulose. The intact granules are insoluble in cold water, but grinding or swelling them in warm water causes them to burst.

The released starch consists of two fractions. About 20 per cent is a water soluble material called amylose. Molecules of amylose are linear chains of several thousand glucose units joined by alpha C-1 to C-4 glycoside bonds. Amylose solutions are actually dispersions of hydrated helical micelles. The majority of the starch is a much higher molecular weight substance, consisting of nearly a million glucose units, and called amylopectin.

Molecules of amylopectin are branched networks built from C-1 to C-4 and C-1 to C-6 glycoside links, and are essentially water insoluble. Glycogen is the glucose storage polymer used by animals. It has a structure similar to amylopectin, but is even more highly branched. The degree of branching in these polysaccharides may be measured by enzymatic or chemical analysis.

Lipids : Lipids are carbon compounds that do not dissolve in water. They are distinguished from other macromolecules by characteristic hydrocarbon chains-long strings of carbon molecules with hydrogens attached. Such chains do not dissolve well in water because they are non-polar. The lipids are a large and diverse group of naturally occurring organic compounds that are related by their solubility in non-polar organic solvents (e.g. ether, chloroform, acetone & benzene) and general insolubility in water. There is great structural variety among the lipids, as will be demonstrated below:

Fatty Acids : The common feature of these lipids is that they are all esters of moderate to long chain fatty acids. Acid or base-catalysed hydrolysis yields the component fatty acid. These long-chain carboxylic acids are generally referred to by their common names, which in most cases reflect their sources.

Natural fatty acids may be saturated or unsaturated, and the saturated acids have higher melting points than unsaturated acids of corresponding size. The higher melting points of the saturated fatty acids reflect the uniform rod-like shape of their molecules. The cis-double bond(s) in the unsaturated fatty acids introduce a kink in their shape, which makes it more difficult to pack their molecules together in a stable repeating array or crystalline lattice. The trans-double bond isomer of oleic acid, known as elaidic acid, has a linear shape and a melting point of 45 °C (32 °C higher than its cis isomer). The shapes of stearic and oleic acids are displayed in the models below.

Two poly unsaturated fatty acids, linoleic and linolenic, are designated "essential" because their absence in the human diet has been associated with health problems, such as scaley skin, stunted growth and increased dehydration. These acids are also precursors to the prostaglandins, a family of physiologically potent lipids present in minute amounts in most body tissues.

Soaps and detergents : Carboxylic acids and salts having alkyl chains longer than eight carbons exhibit unusual behaviour in water due to the presence of both hydrophilic (CO_2) and hydrophobic (alkyl) regions in the same molecule. Such molecules are termed amphiphilic (Gk. amphi = both) or amphipathic. Fatty acids made up of ten or more carbon atoms are nearly insoluble in water, and because of their lower density, float on the surface when mixed with water.

Unlike paraffin or other alkanes, which tend to puddle on the waters surface, these fatty acids spread evenly over an extended water surface, eventually forming a mono molecular layer in which the polar carboxyl groups are hydrogen bonded at the water interface, and the hydrocarbon chains are aligned together away from the water. Substances that accumulate at water surfaces and change the surface properties are called surfactants.

Alkali metal salts of fatty acids are more soluble in water than the acids themselves, and the amphiphilic character of these substances also make them strong surfactants. The most common examples of such compounds are soaps and detergents. The use of such compounds as cleaning agents is facilitated by their surfactant character, which lowers the surface tension of water, allowing it to penetrate and wet a variety of materials. Very small amounts of these surfactants dissolve in water to give a random dispersion of solute molecules.

However, when the concentration is increased an interesting change occurs. The surfactant molecules reversibly assemble into polymolecular aggregates called micelles. By gathering the hydrophobic chains together in the centre of the micelle, disruption of the hydrogen bonded structure of liquid water is minimised, and the polar head groups extend into the surrounding water where they participate in hydrogen bonding. These micelles are often spherical in shape, but may also assume cylindrical and branched forms. The oldest amphiphilic cleaning agent known to humans is soap. Soap is manufactured by the base-catalysed hydrolysis (saponification) of animal fat. Before sodium hydroxide was commercially available, a boiling solution of potassium carbonate leached from wood ashes was used. Soft potassium soaps were then converted to the harder sodium soaps by washing with salt solution. The importance of soap to human civilization is documented by history, but some problems associated with its use have been recognised.

Fats and Oils : The triesters of fatty acids with glycerol (1, 2, 3-trihydroxypropane) compose the class of lipids known as fats and oils. These triglycerides (or triacylglycerols) are found in both plants and animals, and compose one of the major food groups of our diet. Triglycerides that are solid or semi-solid at room temperature are classified as fats, and occur predominantly in animals. Those triglycerides that are liquid are called oils and originate chiefly in plants, although triglycerides from fish are also largely oils.

As might be expected from the properties of the fatty acids, fats have a predominance of saturated fatty acids, and oils are composed largely of unsaturated acids. Thus, the melting points of triglycerides reflect their composition, as shown by the following examples. Natural mixed triglycerides have somewhat lower melting points, the melting point of lard being near 30 ° C, whereas olive oil melts near -6 ° C. Since fats are valued over oils by some Northern European and North American populations, vegetable oils are extensively converted to solid triglycerides (e.g. Crisco) by partial hydrogenation of their unsaturated components. Some of the remaining double bonds are isomerised (to trans) in this operation. These saturated and trans-fatty acid glycerides in the diet have been linked to long-term health issues such as atherosclerosis. Triglycerides having three identical acyl chains, such as tristearin and triolein (above), are called "simple", while those composed of different acyl chains are called "mixed". If the acyl chains at the end hydroxyl groups (1 & 3) of glycerol are different, the centre carbon becomes a chiral centre and enantiomeric configurations must be recognised.

Waxes : Waxes are esters of fatty acids with long chain monohydric alcohols (one hydroxyl group). Natural waxes are often mixtures of such esters, and may also contain hydrocarbons. The formulas for three well known waxes are given below:

spermaceti	:	$CH_3(CH_2)14CO_2\text{-}(CH_2)15GH_3$
beeswax	:	$CH_3(CH_2)24CO_2\text{-}(CH_2)29CH_3$
carnuba wax	:	$CH_3(CH_2)30CO_2\text{-}(CH_2)33CH_3$

Waxes are widely distributed in nature. The leaves and fruits of many plants have waxy coatings, which may protect them from dehydration and small predators. The feathers of birds and the fur of some animals have similar coatings which serve as a water repellent. Carnuba wax is valued for its toughness and water resistance.

Phospholipids : Phospholipids are the main constituents of cell membranes. They resemble the triglycerides in being ester or amide derivatives of glycerol or sphingosine with fatty acids and phosphoric acid. The phosphate moiety of the resulting phosphatidic acid is further esterified with ethanolamine, choline or serine in the phospholipid itself. The following diagram shows the structures of some of these components.

Phospholipid Components

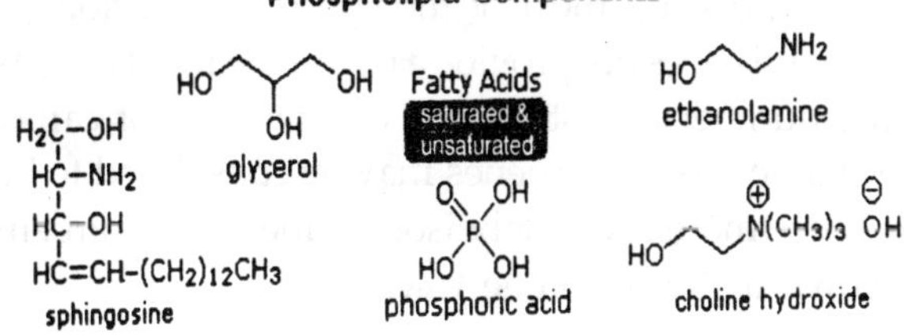

As ionic amphiphiles, phospholipids aggregate or self-assemble when mixed with water, but in a different manner than the soaps and detergents. Because of the two pendant alkyl chains present in phospholipids and the unusual mixed charges in their head groups, micelle formation is unfavourable relative to a bilayer structure.

If a phospholipid is smeared over a small hole in a thin piece of plastic immersed in water, a stable planar bilayer of phospholipid molecules is created at the hole. The polar head groups on the faces of the bilayer contact water, and the hydrophobic alkyl chains form a non-polar interior. The phospholipid molecules can move about in their half the bilayer, but there is a significant energy barrier preventing migration to the other side of the bilayer.

Prostaglandins Thromboxanes and Leukotrienes

The members of this group of structurally related natural hormones have an extraordinary range of biological effects. They can lower gastric secretions, stimulate uterine contractions, lower blood pressure, influence blood clotting and induce asthma-like allergic responses. Because their genesis in body tissues is tied to the metabolism of the essential fatty acid arachadonic acid they are classified as eicosanoids. Many properties of the common drug aspirin result from its effect on the cascade of reactions associated with these hormones.

Terpenes : Compounds classified as terpenes constitute what is arguably the largest and most diverse class of natural products. A majority of these compounds are found only in plants, but some of the larger and more complex terpenes occur in animals. Terpenes incorporating most of the common functional groups are known, so this does not provide a useful means of classification. Instead, the number and structural organisation of carbons is a definitive characteristic. Terpenes may be considered to be made up of isoprene (more accurately isopentane) units, an empirical feature known as the isoprene rule.

Steroids : The important class of lipids called steroids are actually metabolic derivatives of terpenes, but they are customarily treated as a separate group. Steroids may be recognised by their tetracyclic skeleton, consisting of three fused six-membered and one five-membered ring. Steroids are widely distributed in animals, where they are associated with a number of physiological processes. Norethindrone is a synthetic steroid, all the other examples occur naturally.

A common strategy in pharmaceutical chemistry is to take a natural compound, having certain desired biological properties together with undesired side effects, and to modify its structure to enhance the desired characteristics and diminish the undesired. This is sometimes accomplished by trial and error.

Chemical studies of the steroids were very important to our present understanding of the configurations and conformations of six-membered rings. Substituent groups at different sites on the tetracyclic skeleton will have axial or equatorial orientations that are fixed because of the rigid structure of the transfused rings.

This fixed orientation influences chemical reactivity, largely due to the greater steric hindrance of axial groups versus their equatorial isomers. Thus an equatorial hydroxyl group is esterified more rapidly than its axial isomer.

Lipidic vitamins : The essential dietary substances called vitamins are commonly classified as "water soluble" or "fat soluble". Water soluble vitamins, such as vitamin C, are rapidly eliminated from the body and their dietary levels need to be relatively high.

The recommended daily allotment (RDA) of vitamin C is 100 mg, and amounts as large as 2 to 3 g are taken by many people without adverse effects. The lipid soluble vitamins are not as easily eliminated and may accumulate to toxic levels if consumed in large quantity. The RDA for these vitamins are:

Vitamin A 800 μg (upper limit ca. 3000 μg)

Vitamin D 5 to 10 μg (upper limit ca. 2000 μg)

Vitamin E 15 mg (upper limit ca. 1 g)

Vitamin K 110 μg (upper limit not specified)

From this data it is clear that vitamins A and D, while essential to good health in proper amounts, can be very toxic. Vitamin D, for example, is used as a rat poison, and in equal weight is more than 100 times as poisonous as sodium cyanide. These compounds have more than a solubility connection with lipids. Vitamins A is a terpene, and vitamins E and K have long Terpene chains attached to an aromatic moiety. The structure of vitamin D can be described as a steroid. The precursors of vitamins A and D have been identified as the tetraterpene beta-carotene and the steroid ergosterol, respectively.

Proteins

More than half of the organic compounds in cells are proteins, which play an important function in almost every cellular process. Proteins, for example, provide structural support to the cell in the cytoskeleton and make up many of the hormones that send messages around the body. Enzymes, which regulate chemical reactions in the cell, are also proteins. Proteins present in and vital to every living cell.

In the form of skin, hair, callus, cartilage, muscles, tendons and ligaments, proteins hold together, protect, and provide structure to the body of a multicelled organism. In the form of enzymes, hormones, antibodies, and globulins, they catalyse, regulate, and protect the body chemistry. In the form of hemoglobin, myoglobin and various lipoproteins, they effect the transport of oxygen and other substances within an organism. Proteins are generally regarded as beneficial, and are a necessary part of the diet of all animals.

Humans can become seriously ill if they do not eat enough

suitable protein, the disease kwashiorkor being an extreme form of protein deficiency. Protein based antibiotics and vaccines help to fight disease, and we warm and protect our bodies with clothing and shoes that are often protein in nature (e.g. wool, silk and leather). The deadly properties of protein toxins and venoms is less widely appreciated. Botulinum toxin A, from Clostridium botulinum, is regarded as the most powerful poison known. Based on toxicology studies, a teaspoon of this toxin would be sufficient to kill a fifth of the world's population.

The toxins produced by tetanus and diphtheria microorganisms are nearly as poisonous. A list of highly toxic proteins or peptides would also include the venoms of many snakes, and ricin, the toxic protein found in castor beans. Despite the variety of their physiological function and differences in physical properties-silk is a flexible fibre, horn a tough rigid solid, and the enzyme pepsin water soluble crystals—proteins are sufficiently similar in molecular structure to warrant treating them as a single chemical family.

When compared with carbohydrates and lipids, the proteins are obviously different in fundamental composition. The lipids are largely hydrocarbon in nature, generally being 75 to 85 per cent carbon. Carbohydrates are roughly 50 per cent oxygen, and like the lipids, usually have less than 5 per cent nitrogen (often none at all). Proteins and peptides, on the other hand, are composed of 15 to 25 per cent nitrogen and about an equal amount of oxygen. The distinction between proteins and peptides is their size. Peptides are in a sense small proteins, having molecular weights less than 10,000.

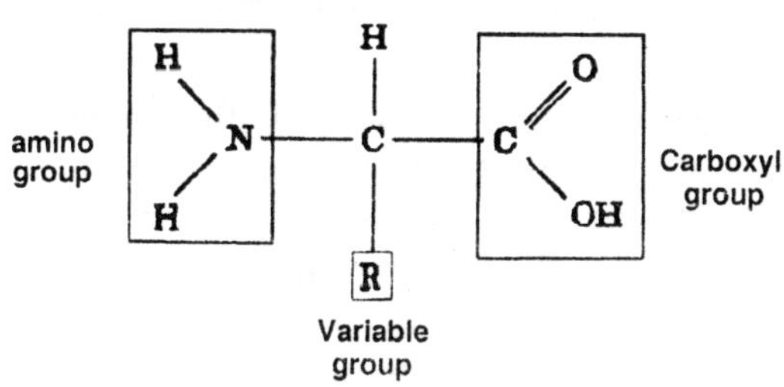

Amino Acids : Proteins are made up of monomers called amino acids. The names of many, but not all, amino acids end in -ine: methionine, lysine, serine, etc. Each amino acid consists of a central carbon atom attached to a set of three designated groups: an atom of hydrogen (-H), an amino group $-NH_2$, and a carboxyl group (-COOH).

It is possible to make an infinite number of amino acids by attaching different compounds to the R position of the central carbon. However, only 20 types of R groups exist in nature, so there are only 20 naturally occurring amino acids.

Natural a-amino acids : Hydrolysis of proteins by boiling aqueous acid or base yields an assortment of small molecules identified as a-aminocarboxylic acids. More than twenty such components have been isolated. Those amino acids having green coloured names are essential diet components, since they are not synthesised by human metabolic processes.

The best food source of these nutrients is protein, but it is important to recognise that not all proteins have equal nutritional value. For example, peanuts have a higher weight content of protein than fish or eggs, but the proportion of essential amino acids in peanut protein is only a third of that from the two other sources. For reasons that will become evident when discussing the structures of proteins and peptides, each amino acid is assigned a one or three letter abbreviation. Some common features of these amino acids should be noted. With the exception of proline, they are all 1°-amines; and with the exception of glycine, they are all chiral.

The configurations of the chiral amino acids are the same when written as a Fischer projection formula, as in the drawing on the right, and this was defined as the L-configuration by Fischer. The R-substituent in this structure is the remaining structural component that varies from one amino acid to another, and in proline R is a three-carbon chain that joins the nitrogen to the alpha-carbon in a five-membered ring. Applying the Cahn-Ingold-Prelog notation, all these natural chiral amino acids, with the exception of cysteine, have an S-configuration.

Many other naturally occurring amino acids exist. Some, such as hydroxylysine and hydroxyproline, are simply functionalised derivatives of a previously described compound. These two amino acids are found only in collagen, a common structural protein. Homoserine and homocysteine are higher homologs of their namesakes. The amino group in beta-alanine has moved to the end of the three-carbon chain. It is a component of pantothenic acid, $HOCH_2C(CH_3)2CH(OH)CONHCH_2CH_2CO_2H$, a member of the vitamin B complex and an essential nutrient. Acetyl co-enzyme A is a pyrophosphorylated derivative of a pantothenic acid amide.

The gamma-amino homolog GABA is a neurotransmitter inhibitor and anti-hypertensive agent. Many unusual amino acids, including D-enantiomers of some common acids, are produced by microorganisms. These include ornithine, which is a component of the antibiotic bacatracin A, and statin, found as part of a pentapeptide that inhibits the action of the digestive enzyme pepsin.

Peptides : If the amine and carboxylic acid functional groups in amino acids join together to form amide bonds, a chain of amino acid units, called a peptide, is formed. A simple tetrapeptide structure is shown in the following diagram. By convention, the amino acid component retaining a free amine group is drawn at the left end (the N-terminus) of the peptide chain, and the amino acid retaining a free carboxylic acid is drawn on the right (the C-terminus). As expected, the free amine and carboxylic acid functions on a peptide chain form a zwitterionic structure at their isoelectric pH.

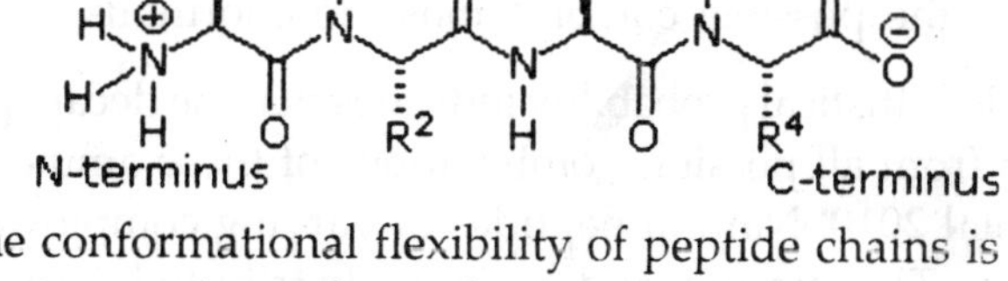

The conformational flexibility of peptide chains is limited chiefly to rotations about the bonds leading to the alpha-carbon atoms. This restriction is due to the rigid nature of the amide (peptide) bond. Nitrogen electron pair delocalisation into the carbonyl group results in significant double bond character between the carbonyl carbon and the nitrogen. This keeps the peptide links relatively planar and resistant to conformational change.

Cyclic Peptides : If the carboxyl function at the C-terminus of a peptide forms a peptide bond with the N-terminal amine group a cyclic peptide is formed. Carboxylate and amine functions on side chains may also combine to form rings. Cyclic peptides are most commonly found in microorganisms, and often incorporate some D-amino acids as well as unusual amino acids such as ornithine (Orn).

Structure of Large Peptides and Proteins : Because the N-terminus of a peptide chain is distinct from the C-terminus, a small peptide composed of different aminoacids may have a several constitutional isomers. For example, a dipeptide made from two different amino acids may have two different structures. Thus, aspartic acid (Asp) and phenylalanine (Phe) may be combined to make Asp-Phe or Phe-Asp, remember that the amino acid on the left is the N-terminus.

The methyl ester of the first dipeptide (structure on the right) is the artificial sweetner aspartame, which is nearly 200 times sweeter than sucrose. Neither of the component amino acids is sweet (Phe is actually bitter), and derivatives of the other dipeptide (Phe-Asp) are not sweet. A tripeptide composed of three different amino acids can be made in 6 different constitutions, and the tetrapeptide (composed of four different amino acids) would have 24 constitutional isomers. When all twenty of the natural amino acids are possible components of a peptide, the possible combinations are enormous.

Simple statistical probability indicates that the decapeptides made up from all possible combinations of these amino acids would total 2010! Natural peptides of varying complexity are abundant. The simple and widely distributed tripeptide glutathione, is interesting because the side-chain carboxyl function of the N-terminal glutamic acid is used for the peptide bond. An N-terminal glutamic acid may also close to a lactam ring, as in the case of TRH (second entry). The different amino acids that make up a peptide or protein, and the order in which they are joined together by peptide bonds is referred to as the primary structure. The various properties of peptides and proteins depend not only on their component amino acids and their bonding sequence in peptide chains, but also on the way in which the peptide chains are stretched, coiled and folded in space. Because of their size, the orientational options open to these macromolecules might seem nearly infinite.

Fortunately, several factors act to narrow the structural options, and it is possible to identify some common structural themes or secondary structures that appear repeatedly in different molecules. These conformational segments are sometimes described by the dihedral angles Φ & Ψ. Most proteins and large peptides do not adopt completely uniform conformations, and full descriptions of their preferred three dimensional arrangements are defined as tertiary structures.

Five factors that influence the conformational equilibria of peptide chains are:

- The planarity of peptide bonds. Conformations are defined by dihedral angles Φ & Ψ.
- Hydrogen bonding of amide carbonyl groups to N-H donors.
- Steric crowding of neighbouring groups.
- Repulsion and attraction of charged groups.
- The hydrophilic and hydrophobic character of substituent groups

Although most proteins and large peptides may have alpha-helix and beta-sheet segments, their tertiary structures may consist of less highly organised turns, strands and coils. Turns reverse the direction of the peptide chain, and are considered to be a third common secondary structure motif.

Approximately a third of all the residues in globular proteins are found in turns. Turns occur chiefly on the protein surface, often incorporate polar and charged residues, and have been classified in three sub-groups. Several factors perturb the organisation of peptide chains. One that has not yet been cited is the structural influence of proline. Unlike the other common amino acids, rotation about the a C-N bond in proline is not possible due to the structural constraint of the five-membered ring. Consequently, the presence of a proline in a peptide chain introduces a bend or kink that disrupts helices or sheets. Also, prolines that are part of a peptide chain have no N-H hydrogen bonding donors to contribute to conformer stabilisation. With the exception of silk fibroin and certain synthetically engineered peptides, significant portions of most proteins adopt conformations that resist simple description or categorisation.

Nucleic Acids : Cells use a class of compounds called nucleic acids to store and use hereditary information. Individual nucleic acid monomers, known as nucleotides, consist of three main units: a nitrogenous base (a compound made with nitrogen), a phosphate group, and a sugar:

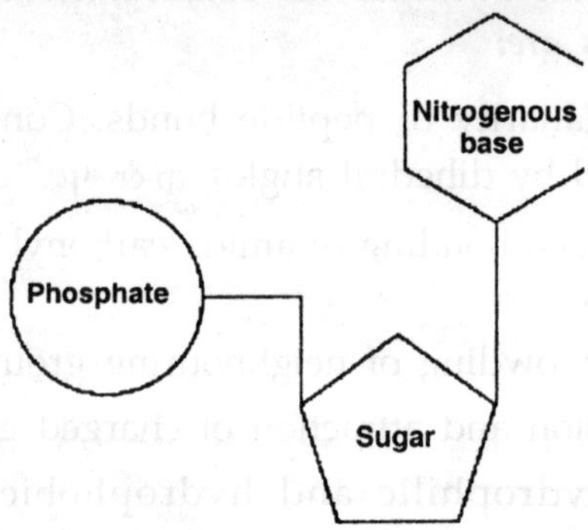

There are two main types of nucleotides, differentiated by their sugars: deoxyribonucleic acid (DNA) and ribonucleic acid (RNA). DNA nucleotides have one less oxygen than RNA nucleotides. The "deoxy" in deoxyribonucleic acid refers to the missing oxygen molecule. In terms of functior, DNA molecules store genetic information for the cell, while RNA molecules carry genetic messages from the DNA in the nucleus to the cytoplasm for use in protein synthesis and other processes. Within both DNA and RNA, there are further subdivisions of nucleotides by nitrogenous bases. For DNA, there are four kinds of nitrogenous bases:

- Adenine (A)
- Guanine (G)
- Cytosine (C)
- Thymine (T)

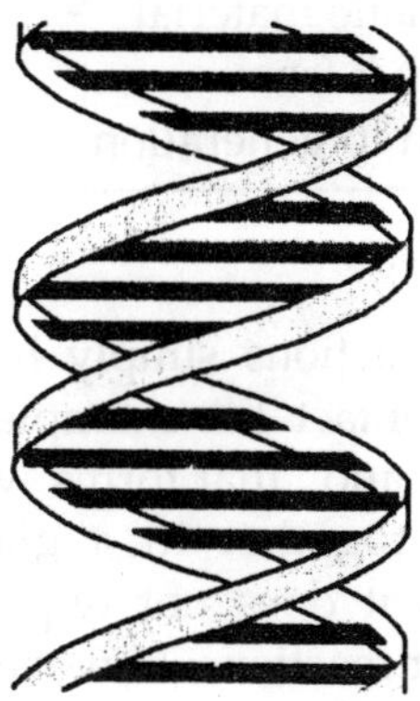

The nitrogenous base of a nucleotide provides it with its chemical identity, so the nucleotides are called by the name of their nitrogenous base. RNA also has four nitrogenous bases. Three-adenine, guanine, and cytosine-are identical to those found in DNA. The fourth, uracil, replaces thymine. Two separate DNA polymers make up each side of the ladder. The sugar and phosphate molecules of the DNA form the vertical supports, while the nitrogenous bases stick out to form the rungs. The rungs attach to each other by hydrogen bonding. The exclusivity of the attachments between nitrogen bases is known as base pairing.

Cytosine	Guanine	DNA and RNA
Adenine	Thymine	DNA only
Adenine	Uracil	RNA only

RNA Structure : Unlike the double-stranded DNA, RNA is single stranded. It looks like a ladder cut down the middle.

This structure of RNA is very important to its functions as a messenger from the DNA in the nucleus to the cytoplasm.

	DNA	*RNA*
Bases	Adenine, guanine, cytosine, thymine	Adenine, guanine, cytosine, uracil
Structure	Double helix	Single helix
Function	Stores genetic material and passes it from generation to generation	Carries messages from the nucleus to the cytoplasm

Enzymes

Some chemical reactions simply happen when the two reactants come into contact. For example, you may be familiar with the bubbly "volcano" that forms when baking soda and vinegar are placed together in a glass. This reaction is spontaneous because it does not require outside energy to force it to occur. Most reactions, however, require energy. For example, the chemical reactions that produce a cake do not take place when baking soda, flour, and the other ingredients of a cake are simply left in a pan on the kitchen counter.

Heat is required to break the existing chemical bonds in the ingredients so that they can undergo chemical reactions and combine with each other in new ways. In the laboratory, chemists use heat to create the activation energy needed to get non-spontaneous reactions started. Animals, however, can't rely on internal Bunsen burners to get their chemical reactions cooking. In order to perform chemical reactions at low temperatures, the body uses special proteins called enzymes, which lower the activation energy necessary for chemical reactions to achievable levels.

Enzymes lower the activation energy by interacting with the substrates, the primary molecules or compounds involved in the reaction. If you think of the activation energy needed for a chemical reaction as a mountain that the reactants have to climb, think of an enzyme as opening up a tunnel through the mountain. Less energy is required to go through the tunnel than to climb all the way up the mountain.

Enzymes are not themselves altered when they help reactions along. Consequently, a single enzyme can be used repeatedly in many reactions. Because enzymes can be used over and over again and because they can act very quickly, a relatively small amount of enzyme is needed to facilitate reactions involving relatively large amounts of material. Each enzyme is designed to fit only the substrates in the reaction that the enzyme is meant to control. The one-to-one correspondence between enzyme and substrate is referred to as specificity. An analogy to a lock and key is useful for understanding the specificity of enzymes. Each enzyme can be thought of as a lock that can interact only with the appropriate key, or substrate. The region of the enzyme that interacts with the substrate is known as the active site.

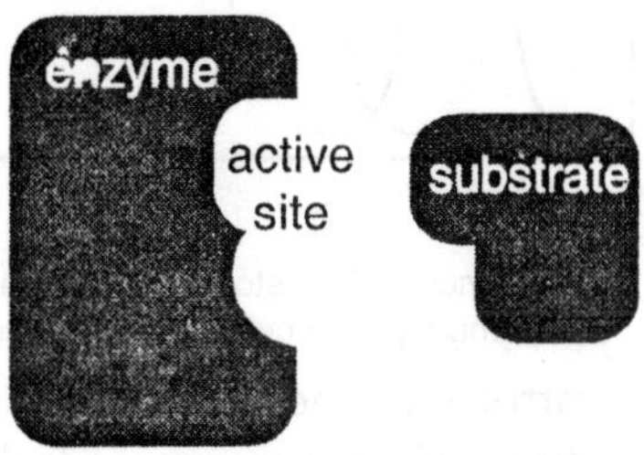

Enzymes help form bonds by holding two substrates near each other in the active site. Compounds can form bonds with each other more easily when they are adjacent than when they are floating around the cell randomly.

Often, enzymes are named for their substrate. The name of the enzyme is the name of the starting material followed by the "-ase." For example, maltase is an enzyme that breaks down maltose, a common sugar.

Affecting Factors of Enzymes : Like all proteins, enzymes have a unique three-dimensional structure that changes under unusual environmental conditions. Enzymes do not function well when their structure is altered.

Temperature and pH : Depending on where it is normally located in the body, an enzyme will have different temperature and pH values at which its structure is most stable. As conditions deviate from this point, the enzyme's ability to help along reactions decreases.

Most enzymes work best near a pH of 7, but some enzymes operate most effectively in a particularly acidic environment, such as the stomach; a neutral environment impairs their function. Likewise, the enzymes of creatures that live at high temperatures, such as bacteria that live in hot springs, do not function properly at human body temperature.

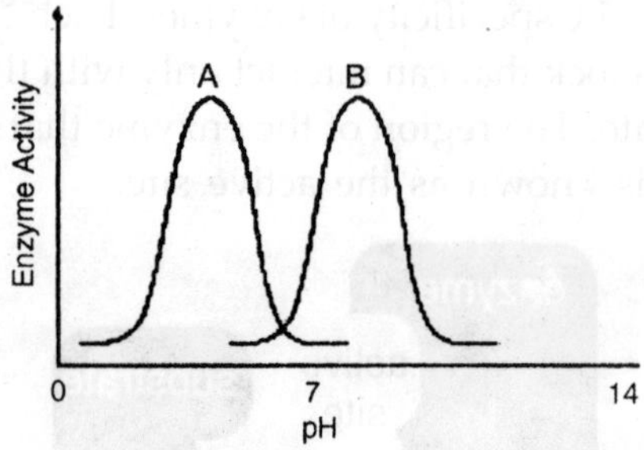

A - optimum pH for stomach enzyme
B- optimum pH for cytoplasm enzyme

In order to control enzyme activity more precisely, the body has developed a number of compounds that turn enzymes on or off and make them work faster or slower. Sometimes these compounds attach to the active site along with the substrate, and sometimes they bind to another site on the enzyme. Activators of enzymes are known as co-factors or co-enzymes. Many vitamins are co-enzymes. Molecules that prevent enzymes from functioning properly are known as inhibitors.

11

Natural Gift

Special Senses

Specialised receptors, or sense organs, are made up of the cells that are usually sensitive to certain types of stimuli, and will not respond to the other types. For example, the eye is specialised for sensing the light rays, but not the sound waves. Receptors are in direct contact with the environment and receive and transmit impressions of what is going on in the environment. In humans, there are special receptors for the perception of sound, light, smell, taste and touch.

Ear as a Sense-organ

The ear is a specialised receptor for detecting sound waves in the surrounding air. The mammalian ear consists of the following three parts:.

Outer Ear: The outer ear or external ear of higher vertebrates is made up of two parts: *(a)* the *pinna,* or skin-covered flap of elastic cartilage and muscles, and *(b)* the external auditory meatus (auditory canal). In humans, the pinna serves the functions of collecting sound waves. In many other mammals, such as the rabbit, the pinna is provided with well developed voluntary muscles and it may be moved so as to collect sound waves more effectively.

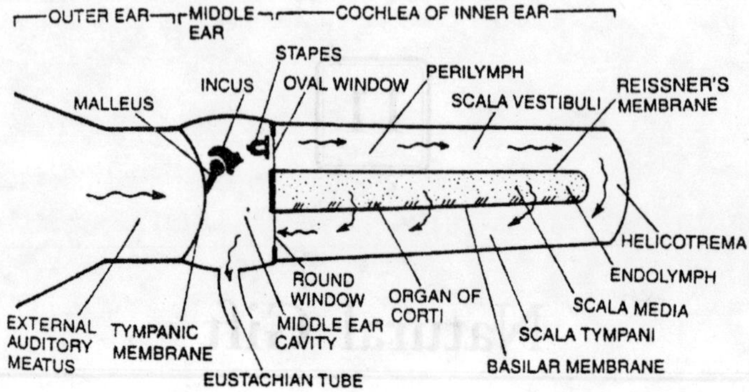

Functional anatomy of the mammalian ear. Arrows indicate the pathway of vibrations.

The auditory canal (about 2.5 cm long) extends inwards and forwards to the ear drum (tympanic membrane). It prevents hard objects from striking the tympanic membrane directly and its walls are lined with hair and wax-secreting glands as a protection against the entrance of foreign objects.

Middle Ear: The middle ear is separated from the outer ear by the *tympanic membrane,* commonly called the *ear drum.* It consists of connective tissue covered with skin on the outside and mucous membrane on the inside. The ear drum is in the shape of a shallow funnel with its apex (umbo) pointing inwards. Movements of the ear drum, set up by the pressure variations caused by the sound waves, are relayed to the three ossicles of the middle ear and thence to the inner ear. The *ossicles* are *malleus, incus* and *stapes.* They are so arranged that the handle of the malleus (hammer) is attached to the inner side of the tympanic membrane. The head of malleus is firmly bound to the incus. The long process of incus passes downwards and parallel to the handle of the malleus and articulates with the head of the stapes (stirrup). The oval base of the stapes fits into the oval window of the cochlea, being surrounded and attached to the margins of the oval window by the annular ligament. The ossicles increase the efficiency of transmission of the sound waves to the inner ear.

To increase the tension of the tympanic membrane for delicate sounds or to counteract under displacement of the membrane, a muscle *tensor tympani* is inserted on the malleus. Contraction of this muscle pulls the malleus and the tympanic membrane inward. A similar muscle *stapedius* is fastened to the stapes and tends to pull the stapes out from the oval window. Like tensor tympani it also serves a protective function and can be reflexly excited by loud sounds.

The middle ear communicates with the pharynx by means of the *eustachian tube,* which regulates the air pressure in the middle ear. Both the middle ear and the eustachian tube are lined with mucous membrane.

Inner Ear: The inner ear consists essentially of two labyrinths, one within the other. The inner one is called the membranous labyrinth which is a closed ectodermal sac filled with endolymph. The *cochlea,* the part involved in hearing, is coiled like a snail's shell, making two-and-a-half turns. The membranous labyrinth is surrounded by a bony labyrinth, which is a hollowed out part of the temporal bone and conforms to the shape and contours of the membranous labyrinth. The space between the two labyrinths is filled up with perilymph.

The cochlea is divided into three longitudinal canals that are separated from each other by two thin membranes called the Reissner's membrane and basilar membrane. These canals become progressively smaller from the base of cochlea to the apex. One of these canals is called the *scala vestibuli* (vestibular canal) whose base is closed by the oval window. The *scala tympani* (tympanic canal) has its base closed by the round window.

At the apex of the cochlea a small opening, the helicotrema, permits continuity between the tympanic canal and the vestibular canal. Between these two canals is the *scala media* (median or cochlear canal). The scala media contains the actual *organ of corti*. The actual sensory receptors in the organ of corti are two types of *hair cells*. There are in all 3500 *inner hair cells* and 20,000 *outer hair cells*. They are innervated by nerve fibres of the cochlear nerve, a division of the eighth nerve. From the upper surface of the hair cells, which are 8-12 μ in diameter, project tiny hair or *stereocilia* 4 μ. long and 0.1 μ thick which pass through a thin dense granular *reticular lamina* and are embedded in the *tectorial membrane*. It is a spiral ribbon attached at one end to the limbus and by one surface and its outer edge to the organ of corti in the region of the Hensen's cells, which lie outside the external hair cells. The tectorial membrane is a stiff gelatinous structure composed of glycoprotein material. Between the two triangular rods of corti lies the tunnel of corti, which is filled with perilymph.

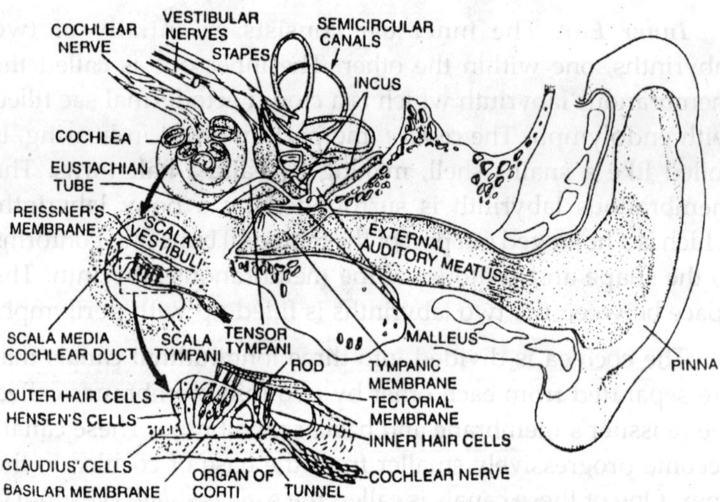

Anatomy of the human ear showing cochlea and organ of corti, etc.

Neural Pathways and Mechanism of Hearing: In hearing, sound waves are received by the external ear and transmitted through the auditory canal to the tympanic membrane which is caused to vibrate. The vibrations are conducted by the chain of ear ossicles to the oval window which transmits them to the fluid in the vestibular and tympanic canals. The vibrations of the endolymph cause the basilar membrane, with its hair cells, to vibrate so that the latter strike the tectorial membrane. This stimulation of hair cells causes them to initiate nerve impulses in the fibres of the auditory nerve, with which they are connected.

There is a spatial organisation of the nerve fibres in the cochlear pathway all the way from the cochlea to the cerebral cortex. And recording of signals in the auditory tracts of the brain stem and in the auditory receptive field of the cerebral cortex shows that specific neurons are activated by specific sound frequencies. Therefore, the major method used by the nervous system to detect different sound frequencies is to determine the positions along the basilar membrane that are most stimulated. This is called *place principle* for determination of sound frequency.

According to the place principle, when sound waves strike the inner ear, the entire basilar membrane is set in vibration by a wave of displacement, which increase the amplitude from the oval window towards the apex of the cochlea. This displacement wave reaches a maximum at the region of the basilar membrane that resonates with the frequency of the incoming sound waves *(von Bekesy,* 1956). The particular hair cells in that region will be stimulated and the impulses conveyed to the fibres of the auditory nerve. Those impulses that are carried by certain fibres of the auditory nerve are interpreted by the hearing centre as particular tones. The loudness of a tone depends on the number of hair cells stimulated, whereas the quality of a tone is produced by the pattern of the hair cells stimulated by sympathetic vibration. The latter characteristic of tone enables one to distinguish between different human voices and different musical instruments even though the tones in each case be of the same pitch and loudness.

Recent findings demonstrate that the sound signals are not transmitted unchanged directly from the ear to the higher levels of the brain; instead, information from the sound signals begins to be dissected form the impulse traffic at levels as low as the cochlear nuclei. Another significant feature of the auditory pathways is that low rates of impulse firing continue even in the absence of sound, with this occuring from the cochlear nerve fibres all the way to the auditory cortex. When the basilar membrane moves towards the scala vestibuli, the impulse traffic increases; when the basilar membrane moves towards the scala tympani, the impulse traffic decreases.

Hard of Hearing

It may be defined as a state in which one is totally or partially unable to hear. The two main types of deafness are: *conductive deafness* and *nerve deafness.* The former may be due to foreign bodies, wax, otitis media which damage the drum and ossicles, or pathological fixation of the stapes in the oval window (otosclerosis). The hearing loss is fairly uniform throughout the frequency range. In patients with conductive deafness air-conducted sounds are less well perceived and bone-conducted sounds may persist longer. In normal subjects the air-conducted sound from the tuning fork is heard longer than that heard by bone conduction when the base of the vibrating fork is placed on the mastoid process (Rinne's test positive). In conductive deafness bone conduction is greater than air conduction (Rinne's test negative).

Nerve deafness may be degenerative, hereditary, toxic, acoustic trauma, meningitis, and acoustic neuroma. In nerve deafness both air-conducted and bone-conducted sounds are less well perceived (Rinne's test).

Means of Equilibrium

Closely connected to the inner ear and forming a part of it are two small sacs, the utriculus and sacculus, and three semi-circular canals . Like the cochlea they are filled with endolymph. They are concerned with the sense of balance and are innervated by the non-acoustic branch of the auditory nerve. The utriculus and sacculus are hollow sacs lined with sensitive hair and each contains a small otolith. The stimulations of the hair cells are interpreted in a certain way with reference to position. The three semi-circular canals are at right angles to each other, one in each plane of space. They are filled with endolymph and at the opening of each canal into the utriculus there is an ampulla which contains sensory hair cells but no otoliths. The hair cells are stimulated by the movement of the endolymph. Thus, these parts of the internal ear are the principal centres of balance.

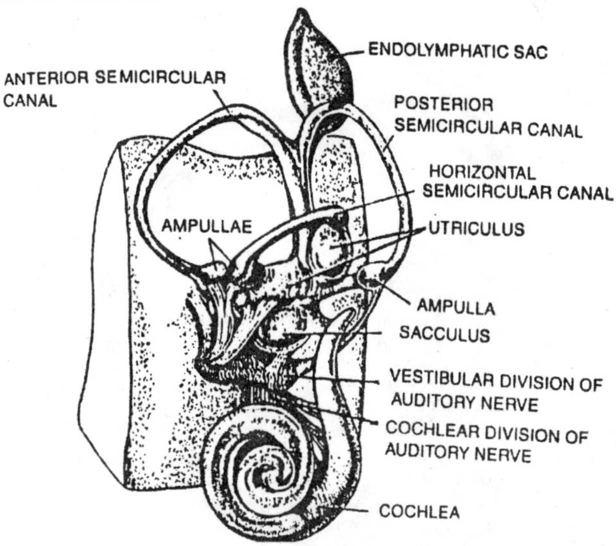

The membranous labyrinth of man.

Eye as a Sense-organ

The human eye is optically equivalent to a photographic camera. It has a lens, a variable aperture (the pupil), and a retina that corresponds to the film. The eye wall is composed of the following three layers

1. The outer layer: The outer layer, known as the *fibrous tunic,* is complete and consists of two regions: (a) the *cornea,* which is the transparent, exposed portion of the tunic, and (b) the *sclera,* the remainder of the tunic to which the extrinsic muscles moving the eye are attached.

The oneness and continuity of the sclera and cornea are, however, more apparent than real. The sclera is a product of a mesodermal envelope that encases the embryonic eye, whereas the cornea is essentially an area of the skin of the head and thus has both ectodermal and mesodermal components. In the adult this head skin is doubled back on itself to form the underlinings of the eyelids, and its epidermal component continues over the cornea as the *conjunctiva.*

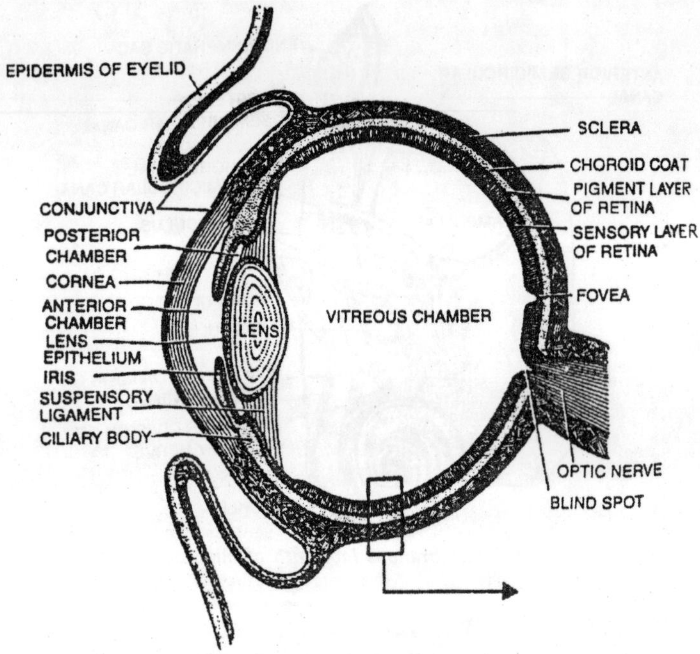

The anatomy of the human eye.

2. *The Middle Layer:* The middle layer is bluish in colour, hence the name uvea. It contains numerous blood vessels and is divided into three regions: (a) the vascular, pigmented *choroid layer,* which is closely adherent to the sclera, (b) the *ciliary body,* which is a relatively thickened, muscular part of the *uvea* at the level of transition between the sclera and cornea, and (c) the *iris.*

From the ciliary body the uvea turns inwards and away

from the sclera and joins with the margin of the innermost layer (retina) to form the *iris,* which is also equipped with muscle. The opening in the centre of the iris is the *pupil.* It can be altered in diameter by the smooth muscle of the iris.

3. *The Innermost Layer:* The innermost layer or the retina abuts the entire extent of the uvea. The part of it in contact with the choroid layer is fairly thick and is the region sensitive to light. The remainder is thin and non-sensory and contributes to the ciliary body and iris. The retina consists of the following layers.

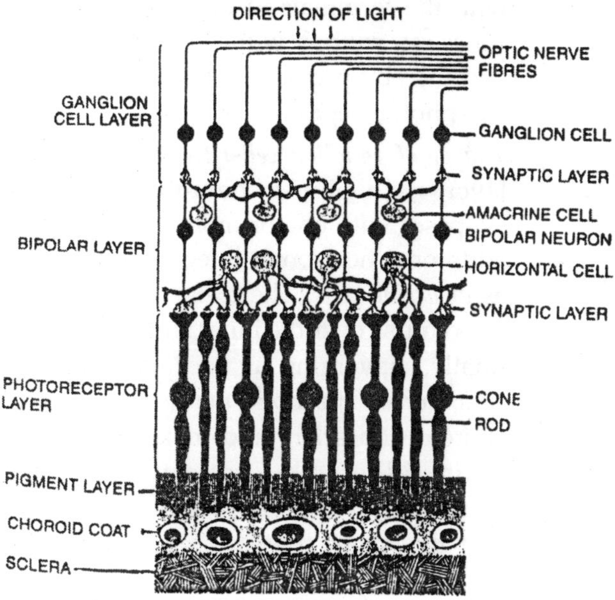

The layers of the retina.

(i) *The Outer Pigmented Layer:* It lies against the choroid coat.

(ii) *The Sensory Layer:* The inner sensory or nervous layer of the retina contains definitive light receptors and a succession of neurons in synaptic series. The cellular components of the nervous retina are, in turn, arranged in three layers. These are:

(a) *a layer of photoreceptors* consisting of long and

narrow rods and shorter, thicker *cones.* In the human retina, there are 125,000,000 rods and 7,000,000 cones. The free ends of rods and cones are oriented towards the outer pigmented layer of the retina.

(b) *a layer of bipolar neurons*: Centrally the rods and cones send processes into the outer synaptic layer where they form synapses with the bipolar neurons. At this level of the bipolar layer there also lie *horizontal cells,* whose processes connect with the terminals of neighbouring rods and cones, as well as with other horizontal cells, and thereby provide horizontal integration of the photoreceptors.

(c) *a layer of ganglion cells:* In the inner synaptic layer, each bipolar neuron terminal forms synpases with one or more ganglion cells and additional horizontal integrative units known as *amacrine cells.* The axons of the ganglionic cells course over the inner surface of the retina, finally converging at one point to form the *optic nerve.* The optic nerve then pierces the choroid and sclera and makes its way to the brain. Where the optic fibres converge, neurosensory cells are necessarily lacking so that this point represents a *blind spot* in the visual field. On the other hand, there is an adjacent area where visual activity is greater than in any other region of the retina. This is *optic fibre centralis,* a highly specialised depression containing many thousands of closely packed cones, but devoid of rods. There are no blood vessels in this area and the ganglionic nerve fibres diverge so that sharply focused light falls directly on the cones.

Thus, the arrangement of the receptors and neurons of the

retina is such that the rods and cones are farthest removed from the light entering the front of the eye. In other words, except for the fovea, light must pass through the layers of ganglionic and bipolar neurons in order to reach the rods and cones. By the same token, stimuli thus set up must pass back through these layers of cells to reach the optic nerve. Because of this reason the retina of vertebrates is said to be "inverted".

Lens and its Application

In the interior of the eye, just behind the iris lies the lens. The lens is clear and glassy in appearance and biconvex in form, with refractive index of 1.42. The lens is held in place by a *suspensory ligament* or zonule composed of fibres radiating from the ciliary body.

The lens system of the eye is composed of four refractive interfaces:

(i) the interface between air and the anterior surface of the cornea,

(ii) the interface between the posterior surface of the cornea and the aqueous humour,

(iii) the interface between the aqueous humour and the anterior surface of the crystalline lens of the eye, and

(iv) the interface between the posterior surface of the lens and the vitreous humour. The refractive index of air is 1; the cornea, 1.38; the aqueous humour, 1.33; the crystalline lens (on the average), 1.40; and the vitreous humour, 1.34. The total refractive power of the eye is about 59 diopters while the total refractive power of the crystalline lens of the eye is only 20 diopters.

Importance of Vision Physiology

Photopigments and Photochemistry of Vision: The photopigment of rods is called *visual purple* or *rhodopsin.* This is a compound composed of a colourless protein opsin, and a yellowish carotenoid, retinaldehyde or *retinene,* a vitamin A derivative. The photopigment of cones is known as *visual violet* or *iodopsin. Wald* (1954) has shown that rhodopsin is bleached by light via intermediaries (lumirhodopsin and metarhodopsin) to the colourless protein *retinene* and *opsin.*

Retinene is reduced to vitamin A by a dehydrogenase working together with $NADH_2$. The equilibrium between vitamin A and retinene lies far over towards vitamin A. If the system were allowed to come into equilibrium retinene would be completely reduced. However, retinene is trapped by opsin to form rhodopsin and is thus continuously removed. Moreover vitamin A, from the blood and from the pigment epithelium, and the continuous oxidation of NAD by the respiratory enzymes assists the vitamin A-retinene equilibrium to move over to the right. The resynthesis of rhodopsin following its breakdown by a light stimulus is responsible for the phenomenon of *dark adaptation.* The importance of vitamin A in the resynthesis and formation of rhodopsin explains how vitamin A deficiency causes night blindness.

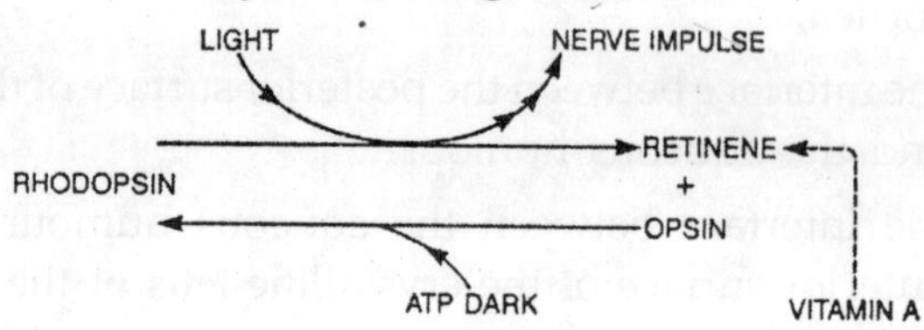

The rhodopsin cycle. The photopigment is rhodopsin in rods and iodopsin in cones. Vitamin A may replenish the supply of retinene and thus indirectly the supply of rhodopsin.

When quanta of light strike the rods or cones, they perform no photochemical work, but trigger the discharge of nerve impulses by the receptor cells. *Wald* (1954) has shown that when rhodopsin is bleached, two to three sulphydryl groups are liberated per molecule of retinene formed. He has suggested that these may initiate the optic nerve impulse either by altering the oxidation-reduction potential or by binding some metallic cation.

Dark Adaptation and Light Adaptation: It is experienced that on passing from a brightly lit room into a dark place, vision at first is very poor but improves as the eye becomes dark-adapted. Thus, *dark adaptation* is the property of the retina which enables the eye to see in the dark. When changing from a bright place to darkness, cone adaptation takes place rapidly, in about 10 minutes. Thereafter, the sensitivity shows a further increase which occurs more slowly but which is of much greater degree. This increase in sensitivity is due to dark adaptation of the rods, for it does not occur in congenitally night-blind (rod function absent) subjects. Thus, if the retina has been exposed to bright light for a long period, complete dark adaptation may take several hours.

In dark adaptation the change in sensitivity of rods depends upon the presence of rhodopsin. As explained earlier, in light the rhodopsin is bleached and is formed only under conditions of low illumination. Thus, the lower the intensity of light, the greater the production of rhodopsin, which increases the sensitivity of rods. As vitamin A is necessary for the formation of retinene and thus indirectly for the supply of rhodopsin, vitamin A deficiency prevents dark adaptation and causes *night blindness.*

Light adaptation occurs within a few minutes when a subject re-enters the light after sitting in darkness. However, the first symptoms are those of dazzle, but the sensations, which may be initially painful, quickly pass off as the eye becomes less sensitive. In addition to adaptation caused by changes in concentrations of rhodopsin, the eye has two other mechanisms for light and dark adaptation. The first of these is a *change in pupillary size.* This can cause adaptation of approximately 30-fold within a fraction of a second, because of changes in the amount of light allowed through the pupillary opening. The other mechanism is *neural adaptation,* involving the neurons in the successive stages of the visual chain in the retina itself and in the brain. This occurs in a fraction of a second although the degree of adaptation is only a fewfold rather than the many thousand-fold that occurs during adaptation of the photochemical system and takes many minutes to hours.

Process of Image-formation

Functionally the eye resembles a photographic camera and the formation of an image on the retina depends on the following:

(i) *Accommodation:* In the eye, the distance between the lens and the retina cannot change. The ability of the eye to focus on objects at varying distances is known as *accommodation* and is due to a mechanism which allows changes of curvature of the anterior surface of the lens. The range of accommodation depends upon the ability of the lens to become more round. In the normal eye the range of accommodation is about 10 dioptres. The distance of an object at which the lens reaches its maximum refractive power is termed the *near point.* It measures about 10 cm for the normal young eye. Thus, such an eye can see objects clearly from infinity down to about 10 cm.

The shape of the lens is controlled by the ciliary body (muscle). In other words, the dioptric strength of the lens depends on the ciliary muscle. When the ciliary muscle is completely relaxed, the dioptric strength of the lens is as weak as it can become. On the other hand, when the ciliary muscle contracts as strongly as possible, the dioptic strength of the lens becomes maximal. Thus, the lens is adjusted to the distance of the object (far object, flat lens; near object, curved lens) by the ciliary muscles until the image becomes focused on the retina.

The ability of lens to change shape progressively decreases with age. As a person grows older, the lens grows larger and thicker and becomes far less elastic, partly because of progressive denaturation of lens proteins. The loss of accommodation is called *presbyopia,* which is due to sclerosis of the lens substance which is unable to adapt itself to a more spherical shape. The power of accomodation decreases from about 14 diopters in a child to less than 2 diopters at age 50 and to about 0 at age 70.

(ii) *Convergence:* In order to perceive only one object when there is an image focused on each retina, the images must fall on corresponding points of each retina. To accomplish this as an object is brought closer to the eyes, the eyes must converge, *i.e.,* turn inward. Such movement is brought about by the extrinsic eye muscles.

(iii) *Change in the Size of the Pupil:* The pupil, corresponding to the diaphragm of a camera, narrows or widens and thereby regulates the amount of light admitted to the eye. The pupillary aperture is controlled by the iris muscles. As these muscles extend, the pupil narrows and less light is admitted. This takes place when bright light falls on the eye. Conversely, when the light intensity is low, the iris muscles contract, thereby enlarging the pupil and permitting more light to enter. The diameter of the pupil can be changed in man from about 1.5 to 9 mm. Since the area of a circle varies with the square of its diameter and because the amount of light admitted by the pupil varies directly with its area, it can be calculated that the amount of light admitted to the eye can be varied about 36 times ($9.0/1.5 = 6$, $6 \times 6 = 36$).

The iris, by altering the aperture of the pupil, not only regulates the amount of light admitted to the eye, but also varies the *depth of focus.* It is the range of distance in which all objects are in focus. In a camera, the depth of focus is increased by decreasing the diaphragm opening. The same is true of the eye. When a lens system has great depth of focus, the retina can be considerably displaced from the focal plane or the lens strength can change considerably and still the image remains in sharp focus, whereas when a lens system has a shallow depth of focus, moving the retina only slightly away from the focal plane causes extreme blurring. The greatest possible depth of focus occurs when the pupil is extremely small.

The iris muscle is innervated by both sympathetic and the

parasympathetic nervous system. The activity of the former increases the pupillary aperture, while the activity of the latter decreases it. The effective stimulus that controls the iris is the intensity of the light striking the retina. The reaction of the iris to a light directed into the eye is termed the *pupillary light reflex.* As the iris of the eye into which the light has been directed changes, so does the iris of the other eye. This is called the *consensual light reflex.*

(iv) *Impulse Interpretation:* An external object is "pictured" on the retina as a series of points, each point corresponding to a rod or a cone. Impulses from these points are transmitted into the brain according to the pattern. All fibres from the left sides of both eyes lead into the left half of the brain and all fibres from the right sides of both eyes lead into the right half of the brain. In each brain hemisphere the fibres from the eyes lead to an optic lobe which contains the visual centres. It can be shown that for each group of rods and cones there exists a corresponding group of interpreter neurons in the visual centres. In other words, the "point picture" of an object on the retina is duplicated more or less faithfully in the optic lobes by impulses from specific rods and cones to their correlated interpreter neurons. In these neurons the impulses register as vision. Each optic lobe thus receives impulses from both eyes and normally the visual interpretations of both lobes give a simple, smoothly superimposed picture of the external world.

The image of an external object is projected on the retina in an inverted position, just as an image is inverted on the film of a camera, but such images are not perceived inverted because the optic centres in the brain give visual experiences correct orientations.

It is also known that visual interpretation in the optic lobes are assisted by the retinal neurons, which carry out a considerable amount of "data processing" before impulses are sent to the brain. For example, it has been found that the retina of a frog generates four distinct sets of impulses to the brain. These four sets of impulses are neatly superimposed in the brain and appear to be interpreted as a single picture. Thus, a frog sees the world in a unique, subjective manner, and a similar conclusion holds for all kinds of light-sensitive animals.

Neural Pathways and Perception of Image: The different neuronal cell types in the retina are: *(a)* the photoreceptors themselves *(rods* and *cones), (b)* the *horizontal cells,* which transmit signals horizontally in the outer plexiform layer from the rods and cones to the bipolar cell dendrites, (c) the *bipolar cells,* which transmit signals from the rods, cones, and horizontal cells to the inner plexiform layer, where they synapse with ganglion cells and amacrine cells, (d) the *amacrine cells,* which transmit signals in two directions, either directly from bipolar cells to ganglion cells or horizontally within the inner plexiform layer between the axons of the bipolar cells, the dendrites of the ganglion cells, and/or other amacrine cells, and (e) the *pinglion cells,* which transmit output signals from the retina through the optic nerve into the brain. A sixth type of neuronal cell in the retina, is the *interplexiform cell,* which transmits signals from the inner plexiform layer to the outer plexiform layer. These signals are all inhibitory and believed to control the lateral spread of visual signals by the horizontal cells in the outer plexiform layer. Their role possibly is to help control the degree of contrast in the visual image.

The neural organisation of the retina clearly shows that the visual pathway from the cones to the ganglion cells is different from the rod pathway. The retina has both an old type of vision based on rod vision and a new type of vision based on cone vision. The neurons and nerve fibres that conduct the visual signals for cone vision are considerably larger than those that conduct the visual signals for rod vision, and the signals are conducted to the brain two to five times as rapidly.

The neurotransmitters released by retinal neurons have not been entirely delineated. Both the rods and the cones release *glutamate* at their synapses with the bipolar cells, and amacrine cells secrete at least eight types of transmitter substances, including acetylcholine, dopamine, glycine, gamma-aminobutyric acid, and indolamine, all of which normally function as inhibitory transmitters. However, the transmitters of the bipolar, horizontal, and interplexiform cells are not clear, but at least some of the synapses of horizontal cells involve electrical transmission rather than chemical transmission.

The transmission of most signals occurs in the retinal neurons by *electrotonic conduction,* not by action potentials. Electrotonic conduction means direct flow of electric current (not action potentials) in the neuronal cytoplasm from the point of excitation to the output synapses. Infact, even in the rods and cones, conduction from their outer segments where the visual signals are generated to the synaptic bodies is by electrotonic conduction. That is, when hyperpolarisation occurs in response to light in the outer segment, almost the same degree of hyperpolarisation is conducted by direct electric current flow to the synaptic body, and no action potential is noticed. Then, when the transmitter from a rod or. cone stimulates a bipolar cell or horizontal cell, once again the signal is transmitted from the input to the output by direct electric current flow, and no action potential occurs. The importance of electrotonic conduction is that it allows *graded conduction* of signal strength. Thus, for the rods and cones, the strength of the hyperpolarising output signal is directly related to the intensity of illumination; the signal is not all-or-none as would be the case for action potential conduction.

It is probable that some of the amacrine cells provide additional lateral inhibition and further enhancement of visual contrast in the inner plexiform layer of the retina as well. It is also known that some bipolar cells depolarise when the rods and cones are excited and others hyperpolarise. Because depolarising and hyperpolarising bipolar cells lie immediately against each other, this provides a mechanism for separating contrast borders in the visual image even when the border lies exactly between the adjacent photoreceptors. The visual pathways can be divided roughly into an *old system* to the midbrain and base of the forebrain, and a *new system* for direct transmission into the *visual cortex.* In humans, the new system is responsible for perception of virtually all aspects of visual form, colours, and other conscious vision. In many primitive mammals, even visual form is detected by the older system, using the superior colliculus in the same manner as the visual cortex is used in mammals.

Vision of Colour

The human eye can recognise about 150 different colours in the visible spectrum (400—750 $m\mu$). The *Young-Helmhotz theory* suggests that there are three types of cones, each perhaps possessing its own characteristic photosensitive substance, one type responding to red, another to green, and third to violet light. These are the primary colours.

The different colour sensations are supposed to be produced by stimulation of various combinations of these three types of receptors. When the three cone types are stimulated equally, a sensation of white light is produced. More recently, evidence has been obtained that there are three different *opsins* in the retina, each in a different cone cell. Their absorption spectra have been measured, and curves were found with peaks at wavelengths of 435, 540, 565 nanometers which are close to the wavelengths of blue, green, and red respectively. Thus, light of a specific wavelength activates one type of cone more than the others, thereby establishing a pattern of impulses different from that evoked by light of a different wavelength. These impulse patterns provide the clue for the appreciation of the colour.

The theory suggests that there is no colour sensitivity in the rods. In order to see colour, there must be relatively high illumination. In dim light everything appears in varying shades of grey because low illumination vision utilises mostly the rods.

Loss of Resolving Power of Eye

Lending support to the Young-Helmhotz theory of colour vision is the clinical finding of patients who are colour blind in one or more of the primary colours. Such findings can be explained on the basis of the existence of three different types of opsins, each specific for a particular colour range. Most commonly, colour-blind people have a deficit either in the red or green range. The terms *protanope, deuteranope* and *tritanope* are used to describe the individuals who have a deficit in the red, green, and blue ranges respectively.

Improper Vision

The following defects may be present:

(i) *Myopia:* If the eyeball is abnormally long, or if the lens in its most elongated state has increased refractive power, parallel rays of light will reach a focus in front of the retina and, therefore, the image will be blurred. Objects close to the eye will be in focus, however, because diverging rays of light from the object will fall in the retina which lies behind the principal focus of the lens. Such a refractive error is termed *myopia,* or more commonly, *nearsightedness.* Glasses with concave lenses of proper refractive power will permit the individual to focus parallel rays of light and to have the normal range of accommodation.

(ii) *Hyperopia:* If the eyeball is too short, or if the lens, in its most relaxed or non-accommodating state, has decreased refractive power, parallel rays of light will reach a virtual focus beyond the eye. This condition is termed *hyperopia,* or more popularly, *farsightedness.* However, such an eye can see distant objects clearly by accommodating.

If there is sufficient accommodation remaining, near objects may also be brought into focus. In this case the need of corrective glasses depends upon two factors: (a) the degree of accommodation required to focus parallel rays of light and *(b)* the remaining range of accommodation. A convex lens is used to correct hyperopia.

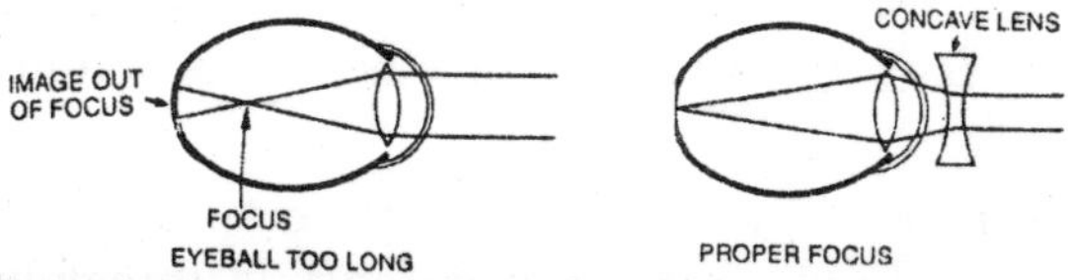

Myopia. The most common cause is an elongated eyeball. The focus point lies in front of the retina. A concave lens that causes parallel rays of light to diverge is required.

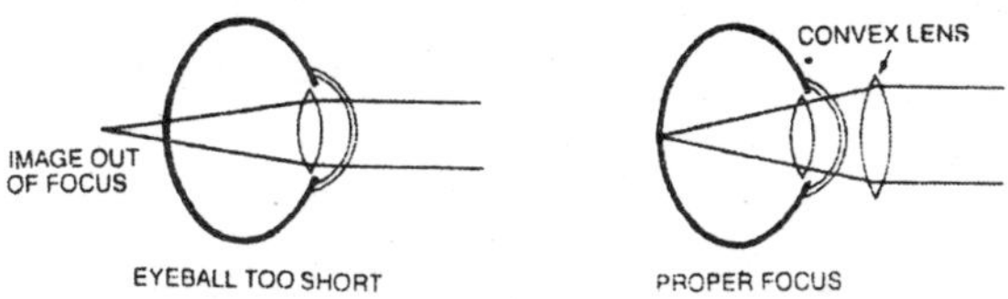

Hyperopia. The most common case is a too short eyeball. The focal point lies beyond the retina. A convex lens that causes parallel rays of light to converge is required.

(iii) *Presbyopia:* With age the lens loses its elasticity and ability to accommodate. Since this decreased range of accommodation is most commonly associated with age, it is termed *presbyopia.* The eyes can no longer accommodate for both near and far vision. Therefore, to see clearly both in distance and nearby, an older person must wear *bifocal glasses* with the upper segment normally focused for far-seeing and the lower segment focused for near-seeing (that is for reading).

(iv) *Astigmatism:* As the name suggests this is an error of vision in which the light rays are not brought to a point focus on the retina. The focus for horizontal rays differs from that for vertical rays. Hence an astigmatic subject looking at a piece of graph paper may focus on the vertical lines and may fail to focus the horizontal lines and vice versa. The defect is most commonly due to a difference in the horizontal and vertical curvatures of the cornea. Occasionally the same abnormality affects the lens. If the curvature is greater along the vertical meridian, astigmatism is said to be 'with the rule.' If the horizontal curvature is greater, the astigmatism is 'against the rule.' To handle astigmatism cylindrical lenses are used which produce refraction in one plane alone.

(v) *Cataract:* The lens of the eye may undergo changes which reduce its transparency. Ultimately the lens becomes cloudy or opaque and is termed a *cataract.* The only treatment is the removal of the opaque lens.

Since the normal lens has a refractive power of about 25 dioptres, glasses with comparable convex lens must be substituted. Because removal of lens removes all power of accommodation, at least two pairs of lenses are required.

(vi) *Strabismus:* The term strabismus means "squint", but it is used to describe a condition in which the image does not fall upon the corresponding point of the two retinas. This is generally due to an abnormality of the extrinsic eye muscles. As a result, two images are seen, a condition termed *diplopia.* It is treated by surgical alteration of the extrinsic muscles, or lenses which will bring the image to bear upon the corresponding points of the two retinas.

(vii) *Argyll Robertson Pupil:* If the pupil fails to change in size in response to light but still responds during accommodation, it is termed as Argyll Robertson Pupil. This is due to the fact that the pathway for the light reflex is different from that for pupillary responses during accommodation.

Functions of Nose and Tongue

Physiologically, smell and taste are related to each other and are generally classified as *visceral senses* because of their close association with gastrointestinal function. Both smell and taste receptors are *chemoreceptors* that are stimulated by molecules in solution in mucous in the nose and saliva in the mouth. The smell receptors are distance receptors *(teleceptors)*. The smell pathways have no relay in the thalamus and there is no neo-cortical projection area for olfaction. The taste pathways pass up the brain stem to the thalamus and project to the post-central gyrus along with those for touch and pressure sensibility from the mouth. However, both senses are strongly tied to primitive emotional and behavioural functions of the nervous system.

What is Smell?

Structure of Olfactory Membrane: The olfactory receptors are located in a specialised part of the nasal mucosa or the *olfactory mucous membrane..* It contains *sustentacular cells,* and *progenitor cells* for the olfactory receptors. Interspersed between these cells are *receptor cells* (olfactory cells). Each olfactory receptor is a neuron having a short thick dendrite with an expanded end called an *olfactory rod* (knob). From each rod, 6 to 12 olfactory hair or cilia project to the surface of the mucous layer. The cilia are unmyelinated processes about 2 μm long and 0.1 μm in diameter. The olfactory membrane also has small *glands of Bowman* which secrete mucous on to the surface of the olfactory membrane.

The axons of the olfactory cells pierce the cribriform plate of the ethmoid bone and enter the olfactory bulbs. In the olfactory bulbs, the axons of the olfactory cells synapse with the primary dendrites of the *mitral cells* to form the *olfactory glomeruli.* The *tufted cells* and the *periglomerular cells* also participate in the formation of glomeruli. The axons of the mitral cells pass posteriorly to the *olfactory cortex* and terminate on the apical dendrites of pyramidal cells in the olfactory cortex. In humans, the olfactory cortex is in the piriform cortex which is bilaterally stimulated by the olfactory stimuli.

Olfaction and its Application

Mechanism for Transmission of Olfactory Signals: The cilia of the olfactory cells first of all respond to the olfactory chemical stimuli. The odorant substance, on coming in contact with the olfactory surface, first diffuses into the mucous that covers the cilia. Then it binds with a *receptor protein* protruding through the caliary membrane. This receptor protein is a long molecule that threads its way through the membrane seven times, folding inward and outward.

The odorant binds with the portion of the receptor that folds to the outside. The inside of the folding receptor, on the other hand, is coupled to a so-called *G-protein,* which is a combination of three subunits (heterotrimeric). On excitation of the receptor, an alpha subunit breaks away from the G-protein and immediately activates adenyl cyclase attached to the inside of the ciliary membrane near the receptor cell body. The activated cyclase in turn converts many molecules of intracellular ATP into cAMP. Finally, this cAMP activates another nearby membrane protein, a gated sodium ion channel that allows large numbers of sodium ions to pour into the receptor cell cytoplasm. The sodium ions contribute positively to the inside of the cell membrane, thus exciting the olfactory nerve cell and transmitting action potentials into the CNS by way of an olfactory nerve.

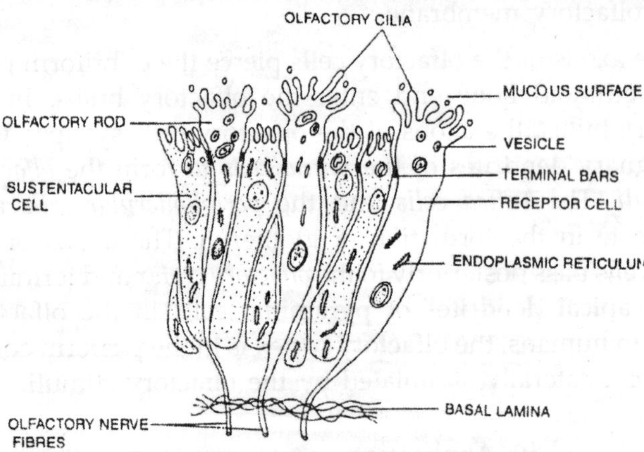

Structure of the olfactory membrane.

The importance of this indirect mechanism for activating olfactory nerves is that it greatly multiplies the excitatory effect of even the weakest odorant substance. In addition to the basic chemical mechanism which stimulates the olfactory cells, several physical factors affect the degree of stimulation. First, only volatile substances that can be sniffed into the nostrils can be smelled. Second, the stimulating substance must be at least slightly water-soluble, as that it can pass through the mucous to reach the olfactory cells. Third, it is helpful to be at least slightly lipid-soluble, because the lipid constituents of the ciliary membrane repel non-lipid-soluble odorants from the membrane receptor proteins.

The membrane potential of unstimulated olfactory cells is about -55 millivolts. At this potential olfactory cells generate continuous action potentials at a very slow rate. Most odorants cause *depolarisation* of the olfactory cell membrane, decreasing the negative potential in the cell from -55 to -30 millivolts or even less. Along with this, the number of action potentials increases to about 20 per second, which is a high rate for the minute olfactory nerve fibres. However, a few odorants hyperpolarise the olfactory cell membrane, thus decreasing, instead of increasing, the nerve firing rate. The rate of olfactory nerve impulses changes approximately in proportion with the logarithm of the stimulus strength, which demonstrates that the olfactory receptors tend to obey principles of *transduction* similar to those of other sensory receptors.

The ability of the olfactory system to discriminate a large number of different odours may be answered that there are many different odorant receptors, and that the cells containing the receptors, with probably only one receptor type in most cells, generate different patterns of neural activity in the brain.

Steps of Olfactory

It is well known that transmission of smell signals from olfactory cell to CNS is through the olfactory bulb. Research suggests that different glomeruli in the olfactory bulbs respond to different odors. Therefore, it is possible that the specific glomeruli that are stimulated are the real clue to the analysis of different olfactory signals transmitted into the CNS. The olfactory tract that enters the brain at the anterior junction between the mesencephalon and cerebrum, is divided into two pathways—one passing medically into the *medial olfactory area* and the other passing laterally into the *lateral olfactory area.* The medial olfactory area is responsible for more primitive responses to olfaction, while lateral olfactory area seems to be important for learning to like or dislike certain foods. However, a third olfactory pathway (that newer pathway) that passes through the thalamus, helps especially in the conscious perception of odour.

Significance of Taste

Taste is mainly a function of *taste buds* in the mouth. The importance of taste lies in the fact that it allows a person to select the desired food. For practical analysis of taste, the receptor capabilities have been arranged into four general categories. They are *sour, salty, sweet,* and *bitter.* The sweet and salty tastes are located principally on the tip of the tongue, the sour taste on the two lateral sides of the tongue, and the bitter taste on the posterior tongue and soft palate.

Framework of Taste Buds

In humans, the taste buds are located in the mucosa of the epiglottis, palate, pharynx and on three types of papillae on the tongue—circumvallate papillae, fungiform papillae, and foliate papillae. Adults have 3000 to 10,000 taste buds.

Each taste bud is an ovoid body measuring 50-70 μ m. It is made up of four types of modified epithelial cells — *basal cells,* type 1 and 2 cells, which are supporting or *sustentacular cells,* and type 3 cells, which are the gustatory receptor cells or *taste cells* that make synaptic connections to sensory nerve fibres. The basal cells arise from the epithelial cells surrounding the taste bud. They differentiate into new receptor cells, and the old receptor cells are continuously replaced in about 10 days.

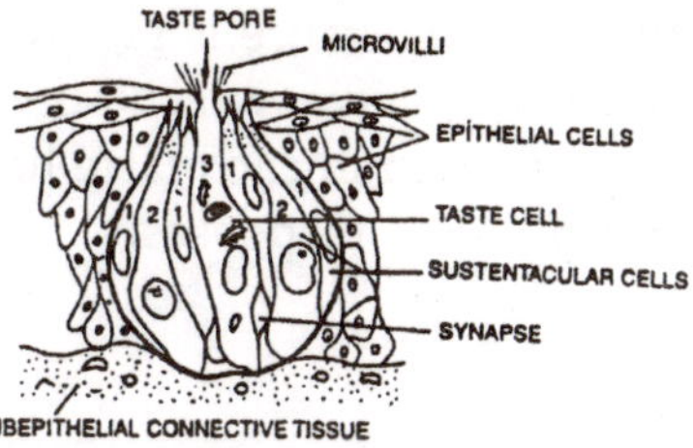

Taste bud, showing type 1,2 and 3 cells.

The outer tips of taste cells are arranged around a minute *taste pore.* From the tip of each cell, several *taste hair* (microvilli) protrude outward into the taste pore. These microvilli provide the receptor surface for taste.

There is a branching terminal network of several *taste nerve fibres* interwoven among the bodies of the taste cells. These fibres are stimulated by the taste receptor cells and some of them invaginate into folds of the taste cell membranes. Many vesicles form beneath the cell membrane near the fibres. These vesicles possibly contain a neurotransmitter substance that is released through the cell membrane to excite the nerve fibre endings in response to taste stimulation.

Taste and its Application

Stimulation of Taste Buds: The membrane of the taste cell, like that of other sensory receptor cells, is negatively charged on the inside with respect to the outside. As soon as the taste hairs are touched by a taste substance, there is partial loss of this negative potential—that is, the taste cell is *depolarised.* The decrease in potential is approximately proportional to the logarithm of concentration of the stimulating substance. This change in potential in the taste cell is the *receptor potential* for taste.

The mechanism by which a stimulating substance reacts with the taste hairs to initiate the receptor potential is by binding the taste chemical to protein receptor molecule(s) that protrude through the taste hair membrane. This in turn opens ion channels, which allow sodium ions to enter and depolarise the cells. Then the taste substance is gradually washed away from the taste hair by the saliva, which removes the stimulus. The type of receptor protein in each taste hair determines the type of taste that will elicit responses.

Steps of Taste

Taste impulses from the anterior two thirds of the tongue pass first into the V cranial nerve, then through the chorda tympanic into the VII nerve, and finally into *tractus solitarius* in the brain stem. Taste impulses from the posterior of the tongue and from other posterior regions of the mouth are transmitted through the IX cranial nerve into the tractus solitarius. Finally, a few taste sensations are transmitted into the tractus solitarius from the base of the tongue and other parts of the pharyngeal region through the X cranial nerve. All taste fibres synapse in the nuclei of the tractus solitarius and send second-order neurons to a small area of the ventral posterior medial nucleus of the thalamus. From the thalamus, third-order neurons are transmitted to taste area in opercula of insula region. Taste does not have a separate cortical projection area in the brain but is represented in the portion of the post-central gyrus in the parietal cortex that subserves cutaneous sensation from the face.

Taste reflexes are integrated in the brain stem to exercise control on the secretion of saliva during the ingestion of food.

Study of Senses Activities

The Eye

The organ of vision consists primarily of the eye-ball sheltered within the *body orbit* and optic nerve. Associated with the eye-ball are various accessories like *extrinsic muscles*, *eyelids* and *tear apparatus*.

The spherical *eye-ball* consists of three coats that enclose the various refractive structures. The outer coat is the tough, fibrous *tunica fibrosa* comprising the transparent *cornea* and the opaque *sclera.* The middle coat is the vascular and pigmented *tunica vasculosa or uvea,* consisting of the *choroid,* the *ciliary body* and the *iris.* The iris ends about a circular opening, the *pupil.* The innermost coat is the *tunica internet,* or *retina.* Its nerve fibres combine posteriorly and leave the eyeball as the optic nerve. A refractive *lens,* held in position by a *suspensory ligament,* lies just behind the pupil and iris.

The vitreous body is a jelly that occupies the large space behind the lens. The space in front of the lens and vitreous body is occupied by watery aqueous humour. The eyelids are movable folds of skin that can cover the front of the eyeball. The lacrimal apparatus secretes *tears* into the *conjunctival sac,* a space between eyelids and eyeball.

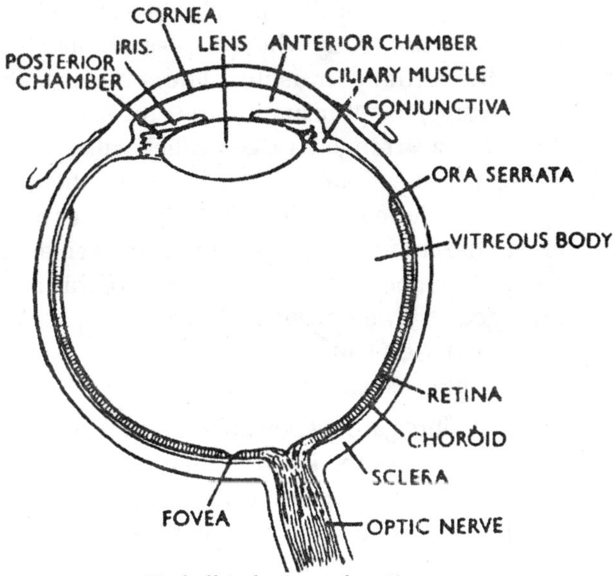

Eyeball in horizontal section.

Fibrous Tunic: The outer coat of eye-ball consists of an opaque sclera and the modified, transparent anterior cornea.

A. *Sclera:* This external coat forms the posterior 4/5 of the eyeball and is continuous anteriorly with the cornea. It ranges from 0.13 to 1.0 mm in thickness. It is fairly expansive and is commonly known as the "white" of the eye. Tendons of the eye muscles insert on this coat.

B. *Cornea:* This is the anterior transparent and non-vascular anterior continuation of the sclera which forms the bulging portion of the eye-ball. From front to back, it consists of 5 well-defined layers.

Vascular Tunic: This middle coat is primarily vascular and pigmented. It is bound closely to the retina but joins the sclera loosely. It has two circular openings: (1) the *pupil* and (2) the one about the optic nerve. From back to front, the middle coat comprises the *choroid, ciliary body* and *iris.*

A. *Choroid:* The choroid is the vascular coat of the eye and consists of four layers.

B. *Ciliary Body:* Anteriorly the choroid is thickened and forms the ciliary body which consists of ciliary processes and the ciliary muscle. The choroid thickens progressively as it nears the iris and here bears about 70 radially arranged ridges called ciliary processes. These processes anchor the fibres of the *suspensory ligament* of the lens. Besides ciliary processes the anterior end of the choroid encloses a prominent ciliary muscle, a mass of smooth muscle. The fibres of this muscle are arranged in three groups: (1) an inner *circular* group near the base of the iris ; (2) a middle *radial* group ; (3) an outer *meridional* group.

The circular fibres are responsible for relaxing the tension on the lens. This allows the lens to bulge by its own elasticity and accommodates the eye for near vision. The radial and meridional fibres are believed to focus the lens for distant vision.

C. *Iris:* This is a further continuation forward of the choroid coat beyond the ciliary body. It serves as a circular adjustable diaphragm that projects into the eye with an aperture, the *pupil,* in its centre. The anterior surface of the iris is covered by a layer of squamous epithelium continuous with that covering the posterior surface of the cornea. The posterior surface of the iris is clothed with an epithelium about 2 layers thick. This is continuous with the epithelium covering the ciliary processes and is derived from the retina. It contains a dark coloured pigment, the *melanin.* The substance of the iris consists of a spongy fibro-elastic tissue which is very vascular. Its upper layer contains many branched pigment cells *(Chromatophores)* that determine the "colour of the eye". A little pigment gives the iris a blue colour, more pigment produces hues of grey and green, while much dark pigment produces a brown iris. In Albinos the pigment as a whole is very scanty and the red colour of the blood in the vessels, combined with the colourless tissue, produces a pink iris.

The iris also contains two groups of smooth muscles, the *circular constrictor* of the pupil near the pupillary margin, and the *radiating dilator* of the pupil situated more externally.

Internal Tunic or Retina: This innermost ocular coat lines the eye-ball ending at the margin of the pupil. Its nervous elements extend only upto the posterior limit of the ciliary body where they end in a wavy line, the *ora serrata.* The extension of the retina forward to the or a serrata is entirely non-nervous. The nervous part of the retina and the optic nerve are essentially parts of the brain. The inner surface of the retina lies against the viterous humour and its outer surface is adjacent to the choroid.

The retina may be divided into ten layers ; from the outside these are: 1. Pigment layer. 2. The photoreceptor layer of rods and cones. 3. Enternal limiting membrane. 4. External nuclear layer (cell bodies of rods and cones). 5. External plexiform layer (synapses of the rods and cones). 6. Internal nuclear layer (bipolar cells). 7. Internal plexiform layer (synapses of the bipolar cells with the ganglion cells). 8. Ganglion cell layer. 9. Nerve fibre layer. 10. Internal limiting membrane.

1. The pigment layer consists of large, cuboidal epithelial cells firmly bound to the Bruch's membrane of the

choroid. The pigment cells bear numerous pigmented, cytoplasmic processes that extend between the rods and cones. Each pigment cell covers about nine or ten of the rods and cones.

2. The photoreceptor layer consists of light-sensitive visual cells of two kinds, *rods* and cones, which correspond to the first neurons in an afferent sensory receptor system. These photoreceptor cells convert light into atomic motion and then into a nerve impulse. These cells are arranged vertically and parallel in a palisade fashion. The rods outnumber the cones in human retina. There are three million cones and a billion rods in a human. The rods are slender cylinders ; cones are shaped like long-necked flasks. Both possess a slender *outer segment* and a thicker *inner segment.* The outer segment of cones is much thicker and conical when seen in section, hence their name.

   (a) Rods are slender, elongated structures, typically 1 μ m in diameter and 40 μ m in length. *The outer segment of a rod is specialised for photoreception.* It contains a stack of about 1,000 discs, which are closed, flattened sacs about 160 Å thick. These membranous structures are densely packed with photoreceptor molecules. The photosensitive molecule in rods is *visual purple* or *rhodopsin,* which consists of *opsin,* a protein, and 11-cis- *retinal,* a prosthetic group. Opsin does not absorb visible light. The colour of rhodopsin and its responsiveness to light depend on the presence of 11-cis-retinal, which is a very effective chromophore. Rhodopsin is synthesised by ribosomes that are attached to the endoplasmic reticulum. A slender cilium joins the outer segment and the *inner segment.* The inner segment of the rod cell, has a large number of mitochondria which are known as the *ellipsoid.* The inner segment generates

ATP at a very rapid rate and is highly active in synthesising proteins. It is contiguous with the nucleus, which is to the synaptic body.

The discs in the outer segment have a life of only ten days and are continually renewed. The visual pigment rhodopsin present in this segment absorbs the light energy and initiates the visual stimulus from the rod cell. These cells are used for peripheral vision and for vision at *low levels of light intensity* (twilight or scotopic vision), for example, the dim light in the interior of a darkened room.

(b) The lamellate plates present in the outer segment of a cone cell do not contain rhodopsin but another visual pigment often called *visual violet* or iodopsin. The cones are thinnest in the fovea where the visual acuity is greatest. The cones are sensitive to *bright light and colour* (photopic vision). Spectrophotometric studies have revealed that these are in fact *three types of cone cells: blue, green,* and *red-absorbing*. The effect of different wavelengths of light on the cones results in the perception of different colours.

3. The rods and cones are separated from one another by radially arranged *glial fibres* also called *Midler's fibres*. These fibres extend through the bulk of the thickness of the retina. Lateral processes of Muller's fibres contact the bases of the rods and cones.

4. The external nuclear layer contains the nucleated cell bodies of rods and cones. Cone nuclei are ovoid and lie just beneath the external limiting membrane whereas rod nuclei are rounder and lie at deeper levels.

5. The external plexiform layer is the region of synapse between the fibres of the rods and cones and the dendrites of the bipolar cells. These latter cells are the second neurons in the afferent chain.

6. The internal nuclear layer contains the nuclei and cell

bodies of *bipolar cells,* of a few large *horizontal cells* and of *'Muller's fibres'.*

7. The internal plexiform layer represents a second synaptic region in the neuron chain. The chief synapses are between axons of bipolar cells and dendrites of ganglion cells.
8. The ganglion cell layer contains the cell bodies and nuclei of large ganglion cells which furnish the third link of afferent neuron chains. The ganglion cells are large neurons with prominent Nissl granules.
9. The axons of ganglion cells form the nerve fibre layer. Here they converge on the papilla to leave the eye-ball as the optic nerve. The retinal elements are absent at the papilla.

   Therefore, this is an area which is insensitive to light and is known as the blind *spot.*
10. The internal limiting membrane forms the innermost layer of the retina. It arises from the expanded ends of *Muller's fibres* that are vertically arranged ependymal cells. These latter cells are supporting elements extending between the two limiting membranes.

An area near the centre of the posterior part of the retina which lies in the direct optical axis is called the 'yellow spot' of *macula lutea.* It is nearly 2 mm in diameter and yellowish when fresh. In the centre of the macula, there is a little depression (0.5 mm wide) called the *fovea centralis.* The number of rod cells in macula is progressively decreased towards its centre, and the fovea only cones are present. The fovea is the region of greatest visual acuity and clearest vision.

Optic Nerve: The *optic nerve fibres* are non-myelinated axons of the ganglion cells of the retina. In the beginning, these fibres pass radially in the nerve-fibre layer of the retina. Near the posterior pole of the eye-ball they turn 90° and enter the optic nerve. The point of exit of the optic nerve is known as the *optic papilla, optic disc* or 'blind spot' because it is insensitive to light.

Just after they leave the eye, the named fibres acquire myelin sheaths. The nerve fibres in each optic nerve total one million.

The optic nerve is an outgrowth of the central nervous system, therefore, the *meninges* (duramater, pia-arachnoid) of the brain cover the optic nerve also. A *retinal artery* and *vein* enter the optic nerve more than midway towards the eye-ball. They proceed to the eye-ball in a strand of connective tissue.

Refracting Media: *Cornea* is the strongest refracting agent of the eye. Behind the cornea, there are four other media that fill the interior of the eye-ball and co-operate in image formation. They are: *crystalline lens, ciliary zonule, vitreous body* and *aqueous humour.*

A. *Crystalline Lens:* The lens is a transparent, elastic, biconvex disc situated just behind the pupil. The lens is surrounded by a homogeneous elastic *capsule.* A single-layered *lens-epithelium* is present on the anterior surface but this layer of cells is absent from the posterior surface of the lens. Rest of the lens is made up of a fibrous *lens substance.* The component *lens fibres* are altered cells in the form of elongate bands which are united by a structureless cementing substance. The central part of the lens substance is older and called the *lens nucleus.* Rest of the lens substance is arranged in concentric *lamella.*

The posterior surface of the lens is more convex than the anterior surface. The two surfaces meet at the rounded off edge called the *equator.*

The major portion of the refraction of light at the eyes does not occur in the lens but at the corneal/air interface. The lens is capable of small changes in shape and thus serves merely to accommodate the eye for distinct vision at varying distances. The lens hardens and becomes opaque in old age due to a deposition of calcium salts. This impairment is termed *cataract* when the lens losses its power of accommodation. This condition is cured by the surgical removal of the lens after which distinct vision is still possible by using biconvex spectacles. However, for near and distant vision lenses of different focal lengths will be required.

B. *Ciliary Zonule:* Attached to the periphery of the lens capsule, both on its anterior and posterior surface is a system of fibres known as the *ciliary zonule.* As the fibres pass outwards they are inserted into the ciliary body. This radially arranged set of fibres serving the lens is a suspensory ligament. Contraction and relaxation of the ciliary muscle changes the thickness of the lens which bends the light rays entering the eye to focus them on the retina. This accommodation is produced by the ciliary muscle and *zonular fibres* of the suspensory ligament.

C. *Vitreous Body:* This is a transparent gelatinous semi-solid jelly that occupies the bulk of the posterior part of the eyeball. It contains fibres, connective tissue cells and some leucocytes. After fixation the jelly appears fibrillar and a *vitreous membrane* can be seen. A *hyaloid canal* runs through the vitreous from the head of the optic nerve to the lens. This marks the site of the hyaloid artery in the embryo. After birth this vessel is not found in the eye.

D. *Aqueous Humour:* The space between the cornea and vitreous body is called the ocular *chamber.* This chamber contains *aqueous humour,* which is a clear, lymph-like fluid. The ocular chamber is incompletely divided by the iris into two parts:

   (1) the *anterior chamber* lying between the cornea, iris and front of the lens, and

   (2) the *posterior chamber* which is a small ring-shaped space bounded by the iris, ciliary body and vitreous body.

Sub Organs

Eyelids: These are movable folds of skin that serve to protect the eye and shut out light. The eyelid is covered externally by skin which is modified on the inner surface as a transparent mucous membrane named the *conjunctiva.* The latter continues, by reflection, over the front of the eye-ball. The conjunctiva is a 4-5 cells thick stratified cuboidal epithelium. It is often the site of inflammations, when the blood vessels supplying it show considerable swelling. This condition is called *conjuctivities.*

The skin on the outer surface of the eyelid consists of a thin epidermis, with small dermal papillae. Sebaceous glands and small hairs are present in this skin. At the edge of the eyelid are present large hair follicles for the *eyelashes.* The latter are coarse hairs with large *sebaceous glands of zeis.* The eyelashes are replaced every 100 to 150 days. Between the hair follicles of eyelashes are large, spiralling sweat glands, the *ciliary glands of Moll* These are provided with large myoepithelial cells.

The shape of the lid is maintained by an internal fibrous plate, the *tarsus.* Numerous sebaceous *tarsalglands of Meibom are* embedded in this plate. Each gland had many alveoli opening into an axial excretory duct. These ducts open externally on the margin of the lid behind the eyelashes. Their oily secretion helps to lubricate the edges of the eyelids so that they do not sick together. It also prevents the entry of water when the eyes are closed. A thin sheet of muscle, the *orbicularis oculi muscle* acts to close the eyelids. The upper lid is raised by the *levator muscle.*

Lacrimal Gland: Each eye is provided with a lacrimal gland which grows out from the upper lateral margin of the conjunctival sac. The lacrimal gland is of the compounds, racemose type. The alveoli have tall cells and a wider lumen, fine cytoplasm of the alveolar cells contains large, pale secretion granules and fat droplets. The secretion provides the clear salty tears that flush the conjectural sac. The tears flow to the inner angle of the eye by 6-12 ducts and are collected by a *lacrimal canaliculus* in each eyelid. By way of the *lacrimal sac* and *naso-lacrimal duct,* they reach the nasal cavity.

Eye Muscles: Four *recti* and two *oblique* voluntary *ocular* *muscles* insert on the sclerotic coat of the eyeball. A single muscle can produce only one type of movement *i.e.* direct lateral whereas all other movements of the eye result from the combined action of at least two muscles.

Vision Physiology

Light is reflected into the eyes by the various objects within the field of vision. White light is actually a combination of all the' colours of the visual spectrum of rainbow, that is, red, orange, yellow, green, blue, indigo and violet. This can be demonstrated by passing white light through a glass prism which refracts or bends the rays of the different colours to a greater or lesser extent. The different colours have different wavelengths, therefore when white light is passed through a prism the different colours take different paths. Red light has the longest wavelength and violet the shortest. This range of colour is called the spectrum of visible light. The spectrum of light is extremely broad but only a small part of it is visible to the human eye.

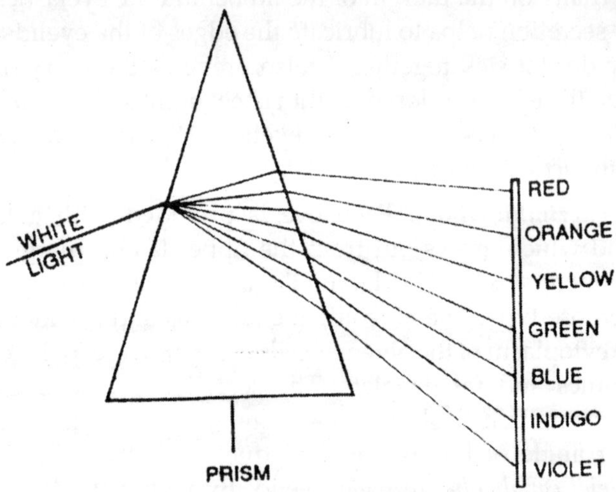

White light broken into the colours of the visible spectrum when passing through a prism.

A specific colour of the visible spectrum is perceived when

only one wavelength of light is *reflected* by an object and all other wavelengths are *absorbed.* For example, when an object appears white all wavelengths are being *reflected,* when it appears *black* all wavelengths are being *absorbed* and it appears *red* when only the *red wavelength is reflected* and all *others are absorbed.* Black and white are not regarded as colours.

The achieve a clear vision light reflected from objects within the visual field is focused on to the retinae of both eyes. Two processes are involved in producing a clear image.

1. Refraction of the light rays.
2. Accommodation of the eyes to light.

Although these may be considered as two separate processes, effective vision is dependent upon their co-ordination.

Movement of Light Rays in Different Mediums

Refraction refers *to bend* or *turn aside.* When light rays pass from a medium of one density to a medium of a different density they are refracted or bent. This principle is used in the eye to focus light on the retina. Several structures which are more dense than air are involved in the refraction of light *in the eye.*

Before reaching the retina light rays pass successively through the cornea, the aqueous humour, the crystalline lens and the vitreous humour.

A lens is a transparent substance that bends rays of light and causes them to converge or diverge. Lens used in spectacles is generally of glass. The ability of a lens to bend light called *refractive power* is expressed in diopters. Focal distance is the distance between a biconvex lens and its *principal focus.* The power of a biconvex lens is expressed in plus diopters and the power of a biconcave lens is termed in minus diopters.

Accommodation of the Eyes to Light: When the eye is at rest the ciliary muscles are relaxed and the lens is thin and stretched, *i.e.,* it has a long focal length. Thus, an eye at rest is adapted for viewing distant objects. When the ciliary muscles contract, the tension in the suspensory ligaments is reduced, thus causing the lens to shrink and become thicker so that it has a short focal length. As the degree of contraction of the ciliary muscles varies, it is possible for the lens to focus on objects at varying distances from the eye. Hence, objects which are very distant *(i.e.* at the limits of visibility) or very near (about 25 cm) can be brought to a focus on the retina. The ability of the eye to bring about the necessary alterations in the focal length of the lens, with regard to the distance of the object from the eye, is known as accommodation.

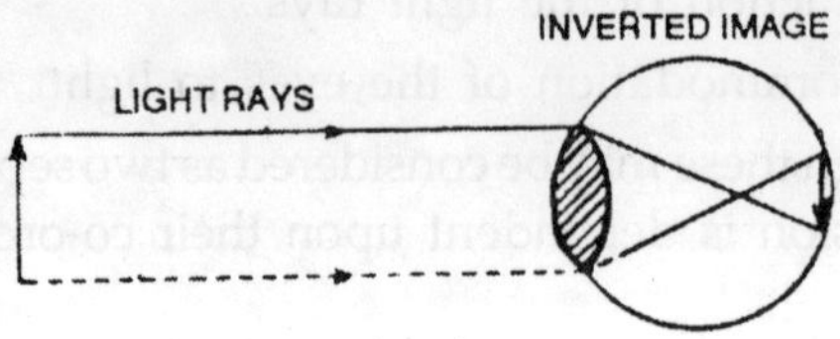

Normal sight.

The normal resting eye (emmetropic) is set for parallel rays travelling from objects 20 feet or more and does not require any accommodation. As objects move nearer to the eyes the rays of light reflected by the object become more divergent and require to be refracted more accurately.

Three factors are involved in accommodation:

A. The size of the pupils.

B. The movement of the eyeballs, called convergence.

C. The increased curvature of lens.

The Size of the Pupils: Pupil size influences accommodation by controlling the amount of light entering the eye. *In a bright light the pupils are constricted. In a dim light they are dilated.*

If the pupils were dilated in a bright light too much light may enter the eye and damage the retina. In a dim light if the pupils were constricted, insufficient light may enter the eye to stimulate the nerve endings in the retina and nothing would be seen.

The iris consists of a layer of circular and a layer of radiating smooth muscle fibres. Contraction of the circular fibres constricts the pupil, and contraction of the radiating fibres dilates it. The size of the pupil is controlled by nerves of the autonomic nervous system. *Sympathetic stimulation dilates the pupils. Parasympathetic stimulation constricts the pupils.*

The Movements of the Eyeballs—convergence: Light rays from objects enter the two eyes at different angles.

It is important that the rays entering should stimulate the corresponding areas of the two retinae. If this is not achieved the individual complains of double vision. Movement of the eyeballs is necessary to ensure this. For near objects both eyeballs move inwards so that their visual axes *converge* on the object to be seen. The nearer the object the greater the convergence. For a distant object less convergence is necessary. The movements of the eye-balls occur due to the contraction of the *extrinsic muscles* of the eyes.

There are six extrinsic muscles situated round the eyeball. Four of these are *straight* and two *oblique*. They are named as follows:

Medial rectus

Lateral rectus

Superior rectus

Inferior rectus

Superior oblique

Inferior oblique.

These muscles have their origin in the bony walls of the orbital cavity and are inserted into the sclera of the eyeball. They are responsible for moving the eye:

The medial rectus rotates the eyeball *inwards.*

The lateral rectus rotates the eyeball *outwards.*

The superior rectus rotates the eyeball *upwards.*

The inferior rectus rotates the eyeball *downwards.*

The superior oblique rotates the eyeball so that the cornea turns in a *downwards and outwards* direction.

The inferior oblique rotates the eyeball so that the cornea turns *upward and outwards.*

The movement of the eyes to look in a particular direction is under voluntary control but the co-ordination of movement of the eyes during convergence and accommodation to near or distant vision is controlled by reflex action.

Increased Curvature of Lens: The lens is a bi-convex, elastic, transparent body suspended by the suspensory ligament from the ciliary muscle. When light rays enter the eye from near objects the function of the lens is to refract these rays on to the macula lutea. When the eyes are accommodated for near vision the *ciliary muscle contracts forward* releasing its pull on the suspensory ligament. As a result there is relaxation of the capsule of the lens. Due to this the lens assumes a more convex form. The anterior surface bulges more than the posterior and the refracting power of the less is increased.

When the eyes are accommodated for distant vision, the *ciliary muscle is relaxed.* The choroid, due to its elastic property, recoils and pulls the suspensory ligament peripherally. As a result, the anterior surface of the lens becomes flattened and the image is correctly focused on the retina.

Formation of Image on the Retina: The mammalian eye works on the same plan as a pin-hole camera. The pupil of the eye corresponds to the diaphragm of the camera, the crystalline lens to the glass lens of the camera and the retina of the eye to the film of the camera. In both the retina and film of the camera, the reaction to light is a chemical one. In the camera, the lens may be moved backward or forward; in the eye, this is accomplished by "accommodation". The inside of the camera is painted black to prevent reflected light from striking the film ; in the eye, the choroid coat, an almost black coat, accomplishes the same purpose.

Light rays from an object travel through the cornea, aqueous humour, lens and vitreous humour. In the process, they are refracted. They finally fall on the retina to form a mirror image which is upside down and smaller than the object. The rods and cones are stimulated and messages are relayed to the visual area of the cerebral cortex through the optic nerve. The messages are then interpreted by the brain cells, which invert the image, and an impression of an upright object is achieved.

Photochemical Changes: George Wald showed that when light falls on the retina *11-cis-retinal group of rhodopsin is isomerised to all-trans-retinal* through several sages. Within a few *picoseconds* of the absorption of a photon, rhodopsin is first converted into a photolytic intermediate, called batho-rhodopsin or pre-lumirhodopsin. In a *nanosecond,* batho-rhodopsin is converted into lumirhodopsin and then to metarhodopsin I in a *microsecond.* The transition from metarhodopsin I to metarhodopsin II takes about a *millisecond.* Metarhodopsin II is hydrolysed in about a *minute* to yield opsin and all-trans-retinal, which diffuses away from the protein.

Rhodopsin (500 nm)

p sec ↓ Light

Bathorhodopsin (543 nm)
(Prelumirhodopsin)

n sec ↓

Lumirhodopsin (497 nm)

μ sec ↓

Metarhodopsin I (480 nm)

m sec ↓

Metarhodopsin II (380 nm)

sec ↓

Opsin + All-*trans*-retinal (380 nm)

Intermediates in the photolysis of rhodopsin. The wavelength of the absorption maximum of each species and the time constant of each transition are given alongwith.

All-trans-retinal is isomerised in the dark to 11-cis-retinal, which associates with opsin to regenerate rhodopsin:

$$\text{All-trans-retinal} \xrightleftharpoons[\text{in the dark}]{\text{Isomerization}} \text{11-cis-retinal}$$

$$\text{11-cis-retinal} + \text{Opsin} \longrightarrow \text{Rhodopsin.}$$

Finally, the ultimate precursors of 11-cis-retinal are the carotenes, which have a very broad biological distribution.

Dark Adaptation: *Visual purple* or rhodopsin is present only in the rods. It is bleached by bright light and when this occurs the rods cannot be stimulated. Rhodopsin is quickly reconstituted when the intensity of light is decreased and an adequate supply of vitamin A is available. When an individual moves from an area of bright light into a dark area, the vision at first is very poor but after sometimes the vision is improved. This is known as dark adaptation. The rate at which dark adaptation takes place is dependent upon the reconstitution of rhodopsin.

Common Defects of the Eye: The common defects of the eye are short-sightedness, long-sightedness and astigmatism. They are mainly caused by excessive eye strain and old age.

Short-Sightedness: This is also known as myopia. This condition occurs when parallel rays of light, *i.e.* rays from a distant object, are brought to a focus in front of the retina and in the vitreous chamber. The defect is the result of the eyeball being too long. The eye is said to be short-sighted and cannot, therefore, focus on distant objects clearly. This defect can be corrected by using a suitable concave spectacle lens which will diverge the light rays entering the eye to the correct extent and bring them to a focus on the retina (divergent-negative power).

Long-Sightedness: This defect maybe due to (i) the eyeball being too short (hypermetropia), *the childhood variety* or *(ii)* the lens and ciliary muscles losing their elasticity with advancing age (presbyopia), the *old age variety* (between 40 and 45 years age group). Long-sighted eyes can only focus on images of distant objects clearly. In both cases, light rays from near objects are brought to a focus behind the retina. This condition can be remedied by the use of a suitable convex spectacle less which will converge the light rays entering the eye to the correct extent and bring them to a focus on the retina (convergent-positive power).

Astigmatism: This is the most common defect of the eye. It is due to the surface of the cornea not being perfectly smooth or spherical, so that rays of light are not brought to sharp points on the retina. This defect is usually present in all eyes to a small extent. In certain cases of astigmatism vertical figures can be seen clearly but horizontal figures appear blurred. Thus, the vertical bars of a window grill will be clear while the horizontal bars will be blurred. In a number of cases, astigmatism may be remedied by the use of cylindrical lenses or by a combination of spherical and cylindrical lenses.

Binocular Vision: Binocular of Stereoscopic vision provides much more accurate assessment of the distance depth, height and width of an object. In this type of vision, each eye 'sees' a scene slightly differently. There is an overlap in the middle but the left eye sees more on the left than can be seen by the right eye and vice-versa. The images from the two eyes are fused in the cerebrum so that one image is perceived.

The Ear

The ear is the 'organ of hearing', as it is receptive to sound waves. It also responds to the effects of gravity and to movements of the head. The ear consists of three parts: (1) The external ear to receive the sound waves, (2) the middle ear to transmit the vibrations, and (3) the inner ear to receive the vibrations and to originate the nerve impulses which are perceived by the central nervous system as sound. The inner ear is also concerned with equilibrium and perception of motion.

With the exception of the pinna, the structures which form the ear are encased within the petrous portion of the temporal bone. The ear is supplied by the *8th cranial nerve.*

The External Ear: The external ear consists of the *auricle, the external auditory maetus* and the *tympanic membrane.*

A. *Auricle (Pinna or Concha):* It is covered by a thin layer of skin containing hairs, large sebaceous glands and small sweat glands. The pinna is shaped and supported by an internal plate of *elastic cartilage.* It is attached to the skull by small vestigial muscles. In old men large stiff hairs often develop at the dorsal edge and lobe of the pinna. It is deeply grooved and ridged. The most prominent outer ridge in humans is called the *helix.* The auricle functions as a sound-collecting appendage.

B. *External Auditory (Acoustic) Maetus:* This is a tube from the auricle to the tympanic membrane. It consists of a cartilaginous *portion* and an *osseous portion.* The outer one-half the tube is supported by elastic cartilage whereas the remainder is a tunnel through the temporal bone. The external acoustic maetus is lined with a continuation of the skin of the auricle. In the outer part of the canal the skin is thick and contains protective hairs, sebaceous glands and large coiled ceruminous glands. These are modified sweat glands of the apocrine type. They discharge a yellowish secretion which mixes with the sebaceous secretion and desquamated cells in the external auditory maetus. Drying of this mixture produces a thick waxy product, the *cerumen.* It has a protective function in that it traps foreign material, such as dust or insects, preventing them from reaching the tympanic membrane.

C. *Tympanic Membrane:* The *ear drum* is a thin semi-transparent membrane set at a strong slant at the deep end of the external maetus. It serves as a common partition between the maetus and tympanic cavity. The membrane is about 0.1 mm thick and consists of three separate types of tissue:

   1. The outer covering of *stratified epithelium* is hairless skin, continuous with that of the maetus.
   2. The middle coat consists of *fibrous tissue.*
   3. The inner lining consists of *cuboidal epithelium* which is continuous with that of the middle ear.

The Middle Ear: The middle ear consists of the tympanic cavity, an air-space traversed by the ear bones, their ligaments and a nerve. It lies within the petrous portion of the temporal bone. The tympanic cavity communicates with the nasopharynx by the *pharyngotympanic* (Eustachian or auditory) *tube.* The cavity is lined by a thin mucous membrane. The epithelium for the greater extent is of squamous type, but near the opening of the Eustachian tube it is columnar and ciliated. Beneath the epithelium is a thin connective tissue which is attached to the periosteum of the surrounding bone.

The medial wall of the tympanic cavity, which separates the middle ear from the inner ear bears an upper oval-shaped vestibular window *fenestra ovalis,* and a lower round cochlear window, *fenestra rotundum.* The oval window is closed by the foot plate of stapes. This window separates the tympanic cavity from the scala vestibuli of the cochlea. The round window separates the tympanic cavity from the scala tympani of the cochlea and is closed by a thin connective tissue membrane often called the *secondary tympanic membrane.* It is covered on the tympanic side by mucous membrane and on the cochlear side by a single layer of squamous cells.

The Auditory Ossicles: Within the middle ear there are three minute bones known as the *auditory ossicles.* The three bones extend across the cavity from the tympanic membrane to the fenestra ovalis like a chain. They form a series of movable joints. The auditory ossicles are: the *malleus,* the *incus* and the *stapes.*

The malleus lies in the lateral aspect of the middle ear. It is shaped like a hummer and possesses a *head, neck* and *handle.* The handle of the malleus is in contact with the medial wall of the tympanic membrane and the head forms a movable joint with the incus.

The incus is an anvil-shaped bone which has a middle Position and articulates with the other two bones. It possesses a *body* and *two processes.* The body articulates with the malleus, the long process with the stapes and the short process is fixed by fibrous tissue to the posterior wall of the cavity. The stapes is shaped like a stirrup presenting a *head, neck, two limbs* and *a base.* The head articulates with the incus and its base fits into the fenestra ovalis.

The three ossicles are maintained in position by fine ligaments. The chain of ear-ossicles acts like a bent lever in transmitting sound vibrations from the tympanic membrane to the endolymph in cochlea.

The Pharyngo-tympanic (Eustachian) Tube: It has a bony and a cartilaginous portion. The bony wall, nearer the tympanic cavity, surrounds 1/3 of the tube, whereas the cartilaginous wall, nearer the pharynx supports 2/3 of the tube. Both portions are lined with a *mucous membrane* whose epithelium near the tympanic cavity is simple columnar and ciliated. Near the pharynx it is pseudostratified and ciliated. The lamina propria contains some *lymphnodules.* The submucosa possesses mixed *sero-mucous glands.*

The Eustachian tube serves to keep the air pressure in the middle ear at atmospheric level, thereby maintaining an equal pressure of air on either side of the ear-drum. This takes place when the tube opens during swallowing.

The Muscles: A *tensor tympani muscle* connects the malleus to the inner wall of the tympanic cavity. This muscle maintains the tone of the tympanic membrane. A *stapedius muscle,* inserted on the inner wall of the tympanic cavity, is attached to the stapes where it joins the incus. This latter muscle is concerned with the articulation of the-ossicles.

The Inner Ear: The inner ear consists of a *bony* or *osseous labyrinth* containing a watery fluid named the *perilymph.* Suspended in the perilymph is a slender *membranous labyrinth* containing a separate fluid, the *endolymph.* The bony labyrinth is hollowed out of the petrous part of the temporal bone. It comprises the *vestibule* that houses the saccule and utricle of the membranous labyrinth, the *semi-circular canals* and the *cochlea.* The bony labyrinth is lined by a layer of flattened cells that covers the periosteum. The membranous labyrinth fits into the bony labyrinth, like a tube within a tube.

Structural Plane of Membranous Labyrinth: It has two divisions. One deals with the static and kinetic senses ; the other, with hearing. The first division comprises five constituent parts:

(1 to 3) Three *semi-circular ducts,* each bearing a swollen *ampulla* at one end. The ducts are set in planes at right angles to each other.

(4) Both ends of the semi-circular canals unite with an ellipsoidal *utricle.*

(5) The utricle joins with a spherical *saccule* by a slender duct. An *endolymphatic duct* also arises from this duct. It extends through the bone and ends in a dilatation, the *endolymphatic sac,* located just outside the duramater.

The hearing sense is served by a tube, the *cochlear duct* coiled spirally like a snail shell. Near its base, it connects with the saccule by the slender *ductus reuniens.* The cochlear duct ends blindly at the apex of the coil. Six areas of the lining epithelium of the membranous labyrinth are specialised as sensory *neuroepithelium.* Innervated by branches of the acoustic nerve, they occur in the utricle, saccule, semi-circular ducts and cochlear duct.

Semicircular Ducts: In cross-section, these ducts are oval and are found eccentrically in the bony canals. On the inner side of the duct, the epithelium is columnar. On the outer side, the thin layer of the connective tissue of the wall is continuous with the strands traversing the perilymphatic space.

Maculae: Both the utricle and saccule bear a thickened, sensory area called the *macula.* It is oval in utricle and neart-shaped in saccule. Their columnar epithelium has two kinds of cells — (1) sensory hair cells and (2) supporting cells.

1. *Sensory hair Cells:* These are shaped like short flasks with a rounded base. Their nuclei lie in an upper row

and the cells do not reach to the basement membrane. The free surface of these cells bears a tuft of sensory *hairs* that are non-motile cilia.

2. *Supporting Cells:* These are slender cells whose nuclei lie at a deep level. The upper portion of each supporting cell is thin and ends at the free surface in a cuticular plate.

The surface of the macula is covered with a gelatinous *otolithic membrane.* The sensory hair tufts project into slender recesses in this gelatinous substance. The recesses are filled with watery endolymph. Towards the upper surface of the membrane are found densely packed minute crystals of calcium carbonate and protein. These particle are called *otoconia.*

The maculae in the utricle and saccule inform of stimuli arising from *linear acceleration* of the body.

Cristae: Each ampulla of the three semi-circular ducts bears a prominence called *crista.* Here the epithelium is columnar and consists of two kinds of cells: (1) the supporting or *sustentacular cells* and (2) the *hair cells.* The supporting cells are narrow in the middle, wide at each end and with basal nuclei. The hair cells are flask-shaped with the nucleus in the expanded portion. They have a tuft of microvilli and one modified cilium or "hair" on the free surface. The hair cells occupy only the upper half of the epithelium.

Covering the surface of the crista is a layer of gelatinous material, the *cupola* into which project the ciliary tufts of the hair cells. The nerve fibres ramify among the hair cells. The cupola is comparable to the otolithic membrane of the macula but it lacks otoconia. The cristae in the ampullae of semi-circular ducts inform about any rotational *movement* involving the head. Any degree of rotation of the head causes movement of the endolymph in the duct leading to a deflection of the cupola. As the processes of the hair cells are embedded in the cupola, any such movement will bend them and stimulate the sensory hair cell. This would initiate nerve impulses leading to the movement of the eyes, head and limbs.

Bony Cochlea: It resembles a snail shell, with a broad base and narrow apex. The osseous cochlea consists of a central axis, the *modiolus* round which a spiral bony canal makes about $2\frac{1}{2}$ turns. A bony shelf projects from the modiolus into the cavity of the shell. This flange is the *spiral osseous lamina.* From its free border a *membranous spiral lamina* or *basilar membrane* continues to the outer wall. At the line of attachment of the basilar membrane to the outer wall of the cochlea the periosteum is greatly thickened, forming the *spiral ligament.* In this way the spiral bony cochlear canal is divided into two compartments, an upper *scala vestibuli* and a lower *scala tymparni.* These are filled with perilymph and communicate with one another at the distal end of the cochlea by the *helicotrema.*

The Membranous Cochlear Duct: The membranous cochlear canal or *scala media* (lying between the scala vestibuli and the scala tympani) is found near the bony cochlear tube. It is filled with *endolymph.* In cross-section, the cochlear duct is triangular and lined throughout by ectodermal epithelium.

The upper surface, or roof, of the cochlear duct is the slanting *Reissner's membrane* (vestibular membrane). Its lower wall consists of the *bony lamina* and *basilar membrane.* The outer or lateral surface blends with thickened periosteum of the bony cochlea. This thickening is the *spiral ligament* The epithelium covering the outer wall is thickened and is very vascular. It is named the *stria vascularis.*

The epithelium resting, on the basal surface or floor of the cochlear duct (formed by bony lamina and the basilar membrane), is highly differentiated. The cells of the outer part are columnar, but those of the inner part are modified into the *spiral organ of Corti,* the receptor for hearing.

Organ of Corti: It extends throughout the two and a half turns of the cochlea. The organs of Corti has a number of special components:

1. *Pillars of Corti:* The pillars of Corti are arranged as two

rows of cells — *inner* and *outer pillars.* Each pillar cell has a broad base and an expanded top or 'head'. The heads articulate together so that the diverging pillars enclose a gable-shaped space, the *inner tunnel of Corti.* A *Nuel's space* lies between the outer pillar and the nearest of the outer hair cells and phalangeal.

2. *Hair Cells:* On either side of the pillars of Corti lie the *hair cells.* These are found is a single row as *inner hair cells* next to the inner pillar, and in four rows as *outer hair cells* outside the outer pillar. Each hair cell possesses a number of microvilli ("hairs") on its outer surface. The tips of the hairs are embedded in the tectorial membrane. The hair cells are short and do not reach the basilar membrane. Their round bases are supported by being enclosed in the phalangeal cells.
3. *Phalangeal Cells (The Supporting Cells Of Dieters):* These enclose the cell bodies of hair cells. They also send long cytoplasmic processes in between them, terminating in an apical umbrella-like expansion. Electron microscope has revealed that the phalangeal cells also enclose afferent and efferent nerve fibres which are passing to the hair cells.
4. *Cells of Hensen:* Outside the outer row of hair cells are 7 or 8 rows of columnar supporting cells, the cells of Hensen. A space between the nearest outer hair cell and Hensens cell is the outer tunnel.
5. *The Tectorial Membrane:* It is a non-nucleated, gelatinous, cuticular membrane which is attached to the osseous spiral lamina. It projects outward peripherally, hanging free over the organ of Corti at the outer limit of the hair cells.

The organ of Corti is receptor for sound. It is responsible for the initiation of nerve impulses which are conveyed to the hearing area in the temporal-lobe of the cerebrum by the auditory nerve (8th cranial nerve). In the brain, these nerve impulses are interpreted as sound.

The *vestibular branch* of the 8th cranial nerve supplied the vestibular part of the inner ear. Its sensory nerve endings are found in the maculae and the cristae. The *cochlear branch* supplies the cochlea.

Activity of Hearing

Every sound that is produced causes vibrations or disturbances in the atmospheric air. These vibrations travel as a succession of waves known as *sound waves.*

Sound waves travel at approximately 335 metres (1088 feet) per second. It is the function of the ear to pick up these vibrations and direct them towards the cochlea, where the organ of Corti is adapted to transmit them, as nerve impulses, by the 8th cranial nerve to the *hearing area* of the cerebral cortex where they are perceived as sound.

Function of the External Ear: The pinna is formed in such a way that the vibrations produced by any sound can be concentrated and directed through the external auditory meatus towards the tympanic membrane. The tympanic membrane vibrates in harmony with the sound waves.

The cerumen secreted by the lining of the external auditory meaturs protects the ear canal.

Function of the Tympanic Cavity or Middle Ear: The vibrations set up in the tympanic membrane are conveyed through the middle ear by a chain of three audited wall of the tympanic membrane and, as it vibrates a corresponding movement of the malleus takes place. The movements of the malleus are transmitted via the incus to the stapes. The foot plate of the stapes fits accurately into the *fenestra vestibuli* The vibrations transferred from the incus to the stapes cause the foot plate of the stapes to rock to and fro in the fenestra vestibuli. This carries the vibrations to the inner ear. The auditory ossicles can only move and transmit the vibrations of the tympanic membrane to the inner ear when air at atmospheric pressure is present in the middle ear. This air enters the middle ear from the nasal part of the pharynx through the Eustachian tube.

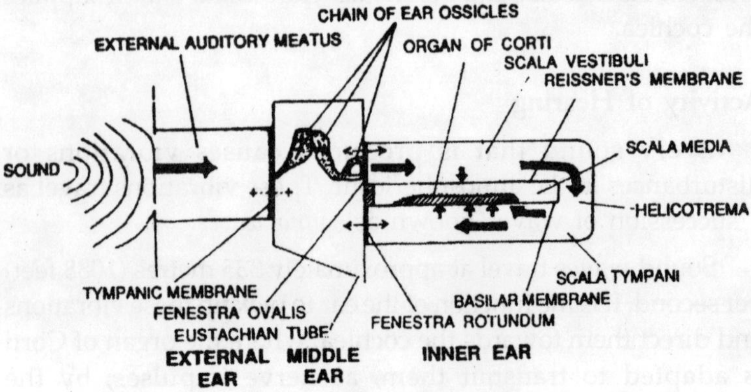

The course of sound waves in the ear.

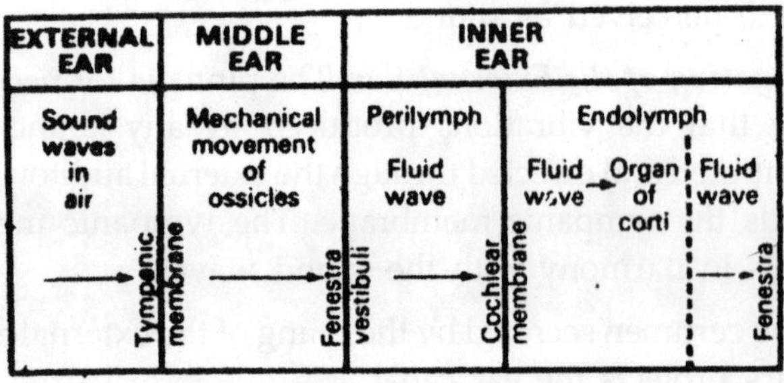

A diagrammatic representation of the transmission of sound through human ear.

Functions of the Inner Ear: On the inner side of the fenestra vestibuli there is perilymph which surrounds the membranous labyrinth. The rocking movement of the stapes sets up wave motion in the perilymph which in turn causes vibration of the membranous labyrinth. The vibrations of the membranous labyrinth stimulate movement of the endolymph, and subsequently the special neuroepithelial cells forming the organ of Corti are stimulated. High-pitched sounds stimulate the cells at the base of the cochlea, and low pitched sounds the cells at the apex.

These nerve impulses are transmitted via the nerve fibres of the *cochlear nerve* which subsequently becomes the

8th cranial nerve. The *vestibulocochlear nerve* transmits the impulses to various nuclei in the pons varolii and midbrain. Some of the nerve fibres pass to the hearing area of the cerebral cortex on the same side and some cross over to the corresponding area in the opposite hemisphere. It is in the hearing area of the cerebral cortex that the nerve impulses are perceived as sound.

Balance: The semi-circular canals, the utricle and the sacculus are of position of head causes movement in the perilymph and endolymph. The sacculus and utriculus contain certain crystals called otoliths which lie near sensitive areas. Nerve fibres form these sensitive areas are embedded in the otoliths. Any movement of the body causes the otoliths to move and pull on the nerve fibres. This produces nerve impulses which are discerned by the brain.

The brain then directs reflexes which tend to return the body to its normal position, each semi-circular canal is situated in a different plane, one being horizontal and two vertical but at right angles to each other. The ampullae of the semi-circular canals contain sensory cells which are stimulated by any movement of fluid in them. This movement stimulates the nerve endings and the hair cells in the utricle, the sacculus and the ampullae. The resultant nerve impulses are then transmitted by the vestibular nerve, which joins the nerve from the cochlea of the inner ear, to form the *vertibulocochlear nerve.* The vestibular branch of the 8th cranial nerve then passes to the *cerebellum.*

Bibliography

Achaya, K.T. : *Indian Dairy Products,* Asia Publishing House, Bangalore, 1974.

Adam, H. : *Food Science,* John Wiley and Sons, New York, 1982.

Addis, Ababa : *Education and Training in Science and Technology in Asia,* UN Economic and Social Council, New York, 1989.

Arbuckle, S. : *Learning of Education,* Allyn and Bacon, Boston, 1965.

Bachman, W. : *Food Chemistry,* Maclaren and Sons Ltd., London, 1955.

Balsdon, J.P. : *Life and Leisure in Ancient Rome,* Bodley Head, London, 1966.

Banerjee, N. : *Indian Females in a Changing Industrial Scenario,* Indo-Dutch Series on Development Alternatives, Sage Publications, New Delhi, 1991.

Cain, M.L. : *Women and Technological Change in Development,* Westview Press, Colorado, 1981.

Carlton, E. : *Foundations of Technical Education,* Prentice Hall, New Jersey, 1964.

Carr, M. : *Appropriate Technology: Two Essays,* Intermediate Technology Development Group, London, 1982.

Crites, J.C. : *Vocational Psychology,* McGraw Hill, New York, 1969.

Donald, E. : *Learning of Technical Education,* Harper Row, New York, 1957.

Dumazedier, J. : *Towards a Society of Leisure,* Free Press, New York, 1967.

Duncan, W.A. : *Engendering School Learning: Science, Attitudes and Achievement among Girls and Boys in Botswana*, University of Stockholm, Stockholm, 1989.

Dyke, S.F. : *The Chemistry of Vitamins*, Interscience Publishers, New York, 1965.

Eley, G. : *Wild Fruits and Nuts*, E.P. Publishing Ltd., Yorkshire, 1976.

Franklin, U.M. : *Will Women Change Technology, or will Technology Change Women?*, The Canadian Research Institute for the Advancement of Women (CRIAW), London, 1985.

Gill, L.S. : *Science, Technology and Gender*, Polity Press in Association with the Open University, Cambridge, London, 1992.

Hanson, J.C. : *Vocational Guidance and Career Development*, MacMillan, New York, 1968.

Helen, Appleton : *Do it Herself: Women and Technical Innovation*, IT Publications, Mumbai, 1995.

Hoyt, K. : *Techniques of Education*, Science Research Associates, New York, 1959.

Hudson, F. : *Development in Food Proteins*, Applied Science Publisher, London, 1982.

James, A.R. : *Education : A System for the Seventies*, Charles E. Marrill Pub-lishing Co., Columbus, 1970.

Jones, A.J. : *Principles of Education in India*, McGraw Hill, New York, 1963.

Kak, Subhash C. : *Indian Physics: Outline of Early History*, ArXiv, London, 2003.

Khosbed, A.W. : *Education Movement in India*, NCERT, New Delhi, 1963.

Krause, M.V. and L.K. Mahann : *Food Nutrition and Diet Therapy*, W.B. Saunders Co., Philadelphia, 1979.

Krouse, H. : *Counselling Technique with Youth*, Charles and Mervil, Ohio, 1960.

Kumari, P. : *Role of Anganwadi Workers in Integrated Child Development Services Programme*, Discovery Publishing House, New Delhi, 1994.

Lagerstrom, B. : *Indian Women in Industry*, Zimbabwe House (Pvt) Ltd., New Delhi, 1984.

Maclean, M.S. : *Counselling of Technical Education*, McGraw Hill, New York, 1955.

Malcolm, S. : *Science and Technology for People*, Tamil Nader State Council for Science and Technology, Madras, 1985.

Marilyn, Carr : *Technology and Women: Conceptual and Empirical Issues*, Allen and Unwin, London, 2000.

Ndunda, M.M.K. : *Women's Resistance to Science Careers in India*, Queen's University, Kingston, 1990.

O'Kelly, Elizabeth : *Simple Technologies for Rural Women in Bangladesh and India*, Kuber Pub., Dhaka, 1978.

Oblsen, M.M. : *Technical Education*, Holt Rinehart and Winston, New York, 1977.

Pacey, A. : *Culture of Technology*, Cambridge, London, 1983.

Pinto, Diana V. : *Education : Theory, Research and Practice*, Rand McNally College Publishing Company, Chicago, 1978.

Potter, N.M. : *Food Science*, The AVI Publishing Co., Connecticut, 1978.

Rahman, A. : *Fatullah Shirazi, a Sixteenth-Century Indian Scientist*, MacMillan, London, 1968.

Rakowski, C.A. : *Women in Non-Traditional Industry: the Case of Steel in Ciudad Guayana*, Venezuela, London, 1985.

Samba, Murthy K.V. : *Science and Technology in Medieval India*, Pratap, Agra, 1982.

Sandler, J. : *The Tech and Tools Book: a Guide to Technologies Women are Using Worldwide*, International Women's Tribune Publications, New Delhi, 1986.

Sharma, P.V. : *History of Medicine in India*, Academic, New Delhi, 1992.

Singh, Harkishan : *Pharmaceutical Education*, Vallabh Prakashan, New Delhi, 1998.

Sponsors : *Proceedings of the Southeast Asian Seminar on Women and Science in Developing Countries*, Kovalevskaia Fund, New York, 1987.

Stadlmare, W.J. and. Cotterill, O.Y. : *Egg Science and Technology*, The AVI Publishing Co., Connecticut, 1973.

Stolte-Heiskanen, V. : *Science: Role of Changing India*, ISSC, New Delhi, 1991.

Subbarayappa, B.V. : *Science and Technological Exchanges between India and Soviet Central Asia*, Educational, New Delhi, 1985.

Theresa, Montini : *Science: Technology and Human Values*, Oxford, New York, 1993.

Tiwari, P.V. : *Kasyapa Samhita or Vrdhajivaka Tantra*, Visvabharati, Varanasi, 1997.

Tolber, E.L. : *Counselling for Career Development*, Houghton, Mifflin, 1974.

Uberoi, Chanchal : *India's Long History of Women in Science*, Physics Today, New Delhi, 1993.

Van, William N. : *Education in the 21st Century*, Houghton Mifflin Co., Boston, 2002.

Veken, M. Van Der : *Women: Technology and Development*, ATOL, Louvain, 1988.

Veldhnis, M.K. : *Food Science and Technology*, The AVI Publishing Co., Connecticut, 1977.

Vijaya Lakshmi, Sharma : *Science in Pre-Modern India*, D.K. Print World, New Delhi, 2001.

Virginia, Hazzard : *Village Technology and Women's Work in Eastern Africa*, UNICEF, Geneva, 1976.

Wardha, B. : *Science and Technology in India*, Centre of Science for Villages, Wardha, 1983.

Whitaker, J.R. : *Principles of Enzymology for Food Science*, Marcel Dekker, New York, 1977.

Winslow, A. : *Science: History*, Duke University Press, London, 1990.

Yukic, T.S. : *Fundamentals of Recreation*, Harper Row, London, 1963.

Zimmerman, J. : *Science in India: Today and Tomorrow*, Praeger Publishers, London, 1983.

Index

P

R

S

T

U

V

W

Z